Dino Klotz

Characterization and Modeling of Electrochemical Energy Conversion Systems by Impedance Techniques

Schriften des Instituts für Werkstoffe der Elektrotechnik,
Karlsruher Institut für Technologie
Band 23

Eine Übersicht über alle bisher in dieser Schriftenreihe erschienenen Bände
finden Sie am Ende des Buchs.

Characterization and Modeling of Electrochemical Energy Conversion Systems by Impedance Techniques

by
Dino Klotz

Dissertation, Karlsruher Institut für Technologie
Fakultät für Elektrotechnik und Informationstechnik, 2012

Impressum

Karlsruher Institut für Technologie (KIT)
KIT Scientific Publishing
Straße am Forum 2
D-76131 Karlsruhe
www.ksp.kit.edu

KIT – Universität des Landes Baden-Württemberg und
nationales Forschungszentrum in der Helmholtz-Gemeinschaft

KIT Scientific Publishing 2012
Print on Demand

ISSN 1868-1603
ISBN 978-3-86644-903-9

Characterization and Modeling of Electrochemical Energy Conversion Systems by Impedance Techniques

Zur Erlangung des akademischen Grades eines

DOKTOR-INGENIEURS

von der Fakultät für

Elektrotechnik und Informationstechnik

des Karlsruher Instituts für Technologie (KIT)

genehmigte

DISSERTATION

von

Dipl.-Ing. Dino Klotz

geb. in: Heilbronn

Tag der mündlichen Prüfung: 19.07.2012

Hauptreferentin: Prof. Dr.-Ing. Ellen Ivers-Tiffée

Korreferenten: Dr. Bernard A. Boukamp

                         Prof. Dr. rer. nat. Michael Siegel

Acknowledgements

First and foremost, I would like to express my gratitude to Prof. Dr.-Ing. Ellen Ivers-Tiffée for her support and confidence in me, that she encouraged me to follow my projects, for providing a laboratory with state-of-the-art measurement equipment, and giving me the possibility to gain further experience by various conference and research trips.

Second, I would like to thank Dr. Bernard A. Boukamp from the Inorganic Materials Science Group (IMS) at University of Twente, Netherlands for his forthright interest in my work and the pleasant and fruitful discussions, and eventually his acceptance to act as secondary reviewer for my thesis.

Just as well I am very thankful for the amazing commitment of Prof. Dr. rer. nat. Michael Siegel, who did not hesitate to accept acting as secondary reviewer from the side of the department. His support is greatly appreciated and most definitely not being taken for granted.

I also want to thank Prof. Dr.-Ing. Volker Krebs and Dr.-Ing. Mathias Kluwe from the Institut für Regelungs- und Steuerungssysteme (IRS), whose support began during my studies and has always been available since then.

Special thanks goes to Ann Call and Kyle Yakal-Kremski from Northwestern University in Chicago and Anna Evans from ETH Zurich for editing the English manuscript.

Further, I would like to thank my group leader Dr.-Ing. André Weber for his support. His ideas together with the discussions and suggestions of my former colleagues Volker Sonn and André Leonide laid the foundation of this work, which never would have been possible without them.

Dr. Tohru Yamamoto of Central Research Institute of Electric Power Industry (CRIEPI) I want to thank for his never ending efforts to provide me a most pleasant and very interesting stay in Yokosuka, Japan. Funding by Japanese Society for the Promotion of Science (JSPS) is also greatfully acknowledged. Another thank you goes to Dr. Koichi Asano and his wife Yukari for giving me a deeper insight into the Japanese culture.

Benjamin Butz I want to thank for his enthusiasm and support concerning the analysis of strange nano-structures that were formed in a Solid Oxide Fuel Cell (SOFC) after the incorrect handling of a measurement device, which eventually led to four papers and four conference contributions.

All my colleagues and teammates at Institut für Werkstoffe der Elektrotechnik (IWE) enabled a very pleasant working atmosphere that made me sometimes feel more at home than in my own apartment. The team has grown too large to mention everybody, but particularly I want to mention Jan Hayd and Jochen Joos being more of friends than of colleagues.

I want to thank my former students and now colleagues Jan Philipp Schmidt and Michael Schönleber, who got convinced to join the IWE during their projects. Jan Philipp designed the first version of the SOFC stack model presented in this thesis and provides me the required plots until now. Michael's enthusiasm to understand, reproduce and finally improve impedance related operations and transformations led to the Time Domain Measurement (TDM) presented in this thesis. Further thanks goes to Jan Richter, whom I co-supervised and who contributed with the constant current measurements, and to Heidrun Köhler, who helped me with the Single Frequency Electrochemical Impedance Spectroscopy (SFEIS).

Frankly, I want to thank my parents, because they always let me do what I wanted and regardless of what I wanted and did were supporting me unquestioned.

Thanks to my former roommates, fellow students and friends, especially Tobias and Miriam Wirsing, Carsten and Melanie Hasberg, Verena Hemm, and Daniela Meijer, for making Karlsruhe a more bearable place to live. Thanks to my clubmates at Tennis Offenau for providing me a clear and sometimes not so clear head – especially Bernd Schulz who supported me throughout my studies and let me go, when I had the foolish idea of writing this thesis.

Last but not least I want to thank Barbara Scherrer for giving me even more reasons to attend conferences by making them and the time after them highly enjoyable.

Dino Klotz

Karlsruhe, June 2012

Zusammenfassung

Die vorliegende Arbeit beschäftigt sich mit der Charakterisierung und Modellierung elektrochemischer Energiewandler. Zwei sehr bekannte Vertreter dieser Art von Systemen sind die Hochtemperatur-Brennstoffzelle (SOFC) und die Lithium-Ionen Batterie (LIB). Ihnen wird allgemein das Potenzial zugeschrieben, künftig zur Erzeugung und Speicherung von elektrischer Energie einen effektiven Beitrag zu leisten, nachdem fossile Brennstoffe und die Atomenergie aus Gründen der CO_2-Belastung, der Sicherheit, der Rohstoffverknappung und nicht zuletzt der Endlagerproblematik langfristig nicht mehr wie im jetzigen Maße für die Grundversorgung mit elektrischer Energie zur Verfügung stehen werden.

Kommerzielle Produkte für beide Systeme sind mittlerweile erhältlich, wobei es sich dabei vor allem bei der SOFC noch eher um Vorserien handelt. Die LIB wird schon in vielen kommerziellen Produkten eingesetzt, jedoch eher im Consumer-Bereich mit kleinen Kapazitäten, die für eine mögliche Energiewende wenig Relevanz haben.

Für den ökonomischen Erfolg dieser Systeme müssen Wirkungsgrad und Herstellungskosten weiter verbessert werden. Dafür sind Einzelzell-Modelle dieser komplexen elektrochemischen Systeme notwendig, die die entstehenden Verluste offenlegen und Verbesserungspotenzial aufdecken. Auf dieser Basis kann dann eine weitere Optimierung stattfinden. Des Weiteren können solche Einzelzell-Modelle auf Systemmodelle hochskaliert werden, um die Wirtschaftlichkeit und die Effizienz im Vergleich zu anderen Systemen zu untersuchen.

Das System, das in dieser Arbeit am ausführlichsten behandelt wird, ist die SOFC – ein System, das aus zwei Gründen sehr interessant für zukünftige Anwendungen im Energiesektor ist. Zum einen ist sie in der Lage, Kohlenwasserstoffe mit einem bisher unerreichten Wirkungsgrad in elektrische Energie umzuwandeln, wobei der Wirkungsgrad im Wasserstoffbetrieb sogar noch höher ist. Zum anderen kann sie ohne großen Aufwand als Elektrolyseur betrieben werden und somit elektrische Energie zur Speicherung in chemische umwandeln. Da dieser Einsatzbereich der SOFC relativ neu ist, müssen die optimalen Betriebsstrategien und Anwendungsmöglichkeiten noch entwickelt werden.

Im ersten Teil dieser Arbeit wird versucht, einen umfassenden Überblick über die allgemeinen, elektrischen, chemischen und thermischen Eigenschaften elektrochemischer Systeme zu geben – mit dem Schwerpunkt auf SOFCs und LIBs.

Eine wichtige Voraussetzung für die Optimierung eines Systems ist die genaue Kenntnis der dabei ablaufenden Prozesse. Es gibt bereits eine Vielzahl an Messverfahren und Charakterisierungsstrategien, die benötigt werden, um die Systeme zu verstehen und um sinnvolle Modelle zu erstellen. Diese werden im zweiten Teil dieser Arbeit vorgestellt. Der hier gegebene Überblick ist breitgefächert, jedoch kann kein Anspruch auf Vollständigkeit gestellt werden. Dafür ist das Feld der elektrochemischen Charakterisierung zu groß.

Aufbauend auf diese Standardverfahren für die Charakterisierung elektrochemischer Systeme werden dann Erweiterungen dieser Verfahren vorgeschlagen, die für eine umfassende objektive Charakterisierung elektrochemischer Systeme benötigt werden.

So ist die wahre Zelltemperatur einer SOFC ein schwer zugänglicher Betriebsparameter, der oft nur ungenau bestimmt wird. Aufgrund der Tatsache, dass die Reaktion, die bei der Wasserstoff-Elektrolyse stattfindet, endotherm ist, hat die Zelltemperatur unter Belastung der Zelle bei der SOEC noch eine viel höhere Relevanz. Aus diesem Grund wird in dieser Arbeit eine Methode entwickelt, mit der die Zelltemperatur ohne zusätzliche Temperatursensoren nur aufgrund des dynamischen Strom-/Spannungsverhaltens bestimmt werden kann. Dies wird erreicht über die Temperaturabhängigkeit des Elektrolytwiderstands. Eine Impedanzmessung bei hoher Frequenz reicht dafür aus. Diese Technik ist nicht komplett neu – sie wurde unter dem Namen Single Frequency Electrochemical Impedance Spectroscopy (SFEIS) schon für einige elektrochemische Systeme angewendet. Jedoch wird in dieser Arbeit erstmals die Anwendung zur Temperaturbestimmung für SOFCs/SOECs gezeigt. Des Weiteren kann aufgrund der hohen Ansprechgeschwindigkeit dieser Technik die Dynamik der Temperaturänderung in der Zelle untersucht werden. Die Methode wurde mittlerweile für LIBs adaptiert, wobei es sogar möglich ist, den Wärmeeintrag über die Modulierung der hochfrequenten Schwingung zu realisieren. Diese Methode wird Electro-Thermal Impedance Spectroscopy (ETIS) genannt. Außerdem konnte SFEIS erfolgreich für die Untersuchung der dynamischen Widerstandsänderung einer LSC-Nano-Kathode aufgrund eines Gaswechsels angewendet werden.

Eine vollständige Charakterisierung eines Systems umfasst den gesamten Frequenzbereich, in dem ein dynamisches Verhalten detektierbar ist. Für SOFCs sind die langsamsten Prozesse noch in einem Bereich, für den die elektrochemische Impedanzspektroskopie (EIS) noch sehr gut geeignet ist. Für LIBs ist dies nicht mehr der Fall, weil hier sehr langsame Prozesse wie die Festkörperdiffusion auftreten, deren Zeitkonstanten im Sub-Millihertz-Bereich liegen. Dies erschwert eine schnelle und präzise Charakterisierung anhand von EIS. Für diesen Zweck werden häufig Zeitbereichsverfahren eingesetzt, deren Ergebnisse mit der Fourier-Transformation ausgewertet werden, um die benötigte Messzeit zu optimieren. Ein sehr leistungsfähiges Zeitbereichsverfahren wird in dieser Arbeit entwickelt, das mit beliebigem Anregungssignal durchgeführt werden kann. In der Literatur existieren einige solcher Verfahren, wobei die meisten davon von einem vordefinierten elektrischen Ersatzschaltbild ausgehen, an das die Zeitbereichsdaten gefittet werden. Dies kann problematisch werden, weil es ein bestimmtes Verhalten des Systems voraussetzt. Außerdem ist es nicht anwendbar für unbekannte Systeme, wo a priori kein gültiges Ersatzschaltbild bekannt ist. Deshalb wird das Impedanzspektrum mit dem hier vorgestellten Verfahren ausschließlich aus den gemessenen Ein-/Ausgangsdaten des Systems berechnet. Es werden keine sonstigen Annahmen getroffen. Im Anschluss an die Herleitung des Verfahrens werden Messergebnisse an einer kommerziellen LIB gezeigt, mit EIS-Messdaten verglichen und diskutiert.

Aber selbst wenn alle relevanten Parameter für die Charakterisierung des Systems im erhaltenen Impedanzspektrum enthalten sind, ist es nicht trivial, diese Parameter aus dem

Spektrum konsistent zu extrahieren. Für technisch relevante Systeme wird dies umso schwieriger im Vergleich zu einfachen Modellsystemen. Allerdings ist es unumgehbar, technisch relevante Systeme zu charakterisieren, weil nur in diesen genau die Prozesse auftreten, die später in einem realistischen Betriebspunkt zu beobachten sind. Auch dafür wurden in der Vergangenheit leistungsfähige Methoden und Algorithmen entwickelt. Aber falls der komplexe, nichtlineare Least-Squares-Fit (CNLS) nicht oder nur in einem lokalen Minimum konvergiert und die Verteilungsfunktion der Relaxationszeiten (DRT) die beteiligten Prozesse nur visualisieren kann, ist es möglich, mit einer Kombination dieser beiden Verfahren – namentlich der Integration der Distribution of Relaxation Times (DRT) in das Gütemaß des Complex Nonlinear Least Squares (CNLS)-Fits – eine robuste und konsistente Parameteridentifikation durchzuführen.

Mit den so erhaltenen Parametern können dann leistungsfähige Modelle parametriert werden. Diese Systemmodelle sind bisher jedoch meist für ein bestimmtes System ausgelegt und verhindern so, dass verschiedene Ansätze objektiv miteinander verglichen werden können. Deshalb wird im letzten Teil dieser Arbeit ein Modell vorgeschlagen, das auf dem statischen Arbeitspunkt und der Impedanz in diesem Arbeitspunkt basiert. Auf Basis eines erprobten Ersatzschaltbild-Modells für kleine SOFC Einzelzellen wird in dieser Arbeit ein Systemmodell hergeleitet. Es besteht aus einem Modell mit verteilten Parametern, die die lokalen Betriebsparameter an jedem Ort im System widerspiegeln, was wichtig ist, weil das elektrochemische Verhalten des Systems an diesem Ort nämlich sehr stark von diesen Betriebsparametern abhängt.

Das Modell liefert eine Vielzahl an Informationen über das modellierte System. Es können verschiedene Modelle mit unterschiedlichem Zweck daraus abgeleitet werden, die mit statischen Betriebsparametern und den entsprechenden Impedanzdaten parametriert werden können. Das wichtigste Ergebnis hieraus ist ein statisches System-Modell, dessen Genauigkeit anhand von Messungen demonstriert wird. So wurde das Modell mit Messdaten von kleinen SOFC Einzelzellen ($A_1 = 1\,\mathrm{cm}^2$) parametriert und kann das statische Verhalten einer großen SOFC Einzelzelle ($A_{16} = 16\,\mathrm{cm}^2$) sehr gut wiedergeben.

Des Weiteren wurde ein detailliertes elektrochemisches System-Modell erstellt, das die Einzelverluste innerhalb des Systems den zuvor an der Einzelzelle identifizierten Prozessen zuordnen kann.

Zuletzt wurde die Basis für ein dynamisches System-Modell gelegt. Im derzeitigen Status ist die Genauigkeit nicht ausreichend, um die gemessene Impedanz einer großen Zelle zu berechnen. Dafür sind weitere Messungen und neue Modellierungsansätze notwendig. Nichtsdestotrotz konnte das Modell schon helfen, die Gasumsatz-Impedanz (GCI) als zusätzlichen Prozess in einem großen System in Messungen zu identifizieren und als zusätzlichen Prozess für das System-Modell zu quantifizieren. Dies lieferte bereits wichtige Hinweise für die Schwierigkeiten, die entstehen, wenn das Verhalten einer Einzelzelle auf ein technisch relevantes System hochskaliert werden soll. Die Erkenntnisse hierbei sind weiterhin nicht auf die SOFC beschränkt und können auf andere elektrochemische Systeme übertragen werden. Der gewählte Modellansatz für alle in dieser Arbeit hergeleiteten Modelle basiert auf statischen und dynamischen Strom-/Spannungsmessungen und lässt sich dadurch für jegliche Art von elektrochemischen Systemen anwenden. Auf Basis eines auf diese Weise erstellten System-Modells ist es nun wiederum möglich, ein Diagnose-Modell für ein Prozessleitsystem zu erstellen.

Obwohl das Reverse Current Treatment (RCT) in dieser Arbeit nicht den ihm gebührenden Stellenwert bekommen hat, sei zu erwähnen, dass die Methode – nämlich durch einen Strompuls in negativer Stromrichtung – bei eine SOFC Einzelzelle in situ in der Lage ist, eine sehr leistungsfähige Nano-Schicht zwischen Elektrolyt und Anode zu generieren. Die Resultate dieser Untersuchungen wurden bereits diverse Male publiziert und präsentiert und sind jeweils auf reges Interesse gestoßen. Momentan wird dieses Thema weiterfolgt und es wurden bereits weitere vielversprechende Ergebnisse für die Leistungsverbesserung erzielt.

All diese Ansätze und Resultate haben das Ziel, elektrochemische Energiewandler in der Zukunft besser zu verstehen und dadurch eine weitere Optimierung zu ermöglichen. Sie sollen weiterhin helfen, objektive Werte für die Leistungsfähigkeit, Effizienz und Wirtschaftlichkeit zu generieren. Belastbare Antworten auf diese Fragen werden ein wichtiger Faktor für die Planung der zukünftigen Struktur der Erzeugung und Speicherung von elektrischer Energie sein.

Contents

1. Introduction

The projected rate of depletion of fossil fuels, as well as the climate change related to their combustion and the correlated ecological footprint of mankind have become widely discussed issues. Nuclear power plants have appeared to be a reliable source of electrical energy without emissions. However, the recent incident at the Fukushima Daiichi Power Plant in Japan has once again demonstrated their threatening danger. In order to meet the rising future demand for energy, new forms of electrical energy production and storage must be engineered.

For this purpose, research in the field of electrochemical systems has been intensified in recent years. The direct transformation of chemical energy into electrical energy and vice versa yields the potential of high efficiencies, fuels flexibility, as well as fast and efficient storage of electrical energy. Electrochemical energy converting systems are versatile and are capable of meeting energy needs for a variety of applications ranging from consumer products to medium-sized power plants.

Current research efforts have been focused on optimization of the applied materials systems and the identification of basic reaction steps in electrochemical devices. A large number of models have been developed from these research findings, however, the versatility of most of these proposed models is small and they are not appropriate for systems other than those that they were designed for. This is generally due to the requirement of undesired simplifications or generalizations, or alternatively from overly complex derivations with a high number of parameters and interdependencies that complicate their handling.

The impartial electrochemical characterization – in order to compare systems and approaches and to estimate their long term and maximum power capabilities – remains the important discipline, that will answer the question, whether a new system can be operated economically sound or not. For this task independent physical quantities have to be deduced from measurements through elaborate models to rate the performance of the system under test, because these quantities are not directly accessible for electrochemical systems.

Goal of this Thesis

The goal of this thesis is to provide a comprehensive overview of advanced techniques for electrochemical characterization of fuel cells and batteries with focus on Electrochemical Impedance Spectroscopy (EIS). Beyond that, amendments to standard EIS analysis are suggested, which aim to improve measurement time, parameter identification, and tracking of evolving parameters. Results will be demonstrated with the help of actual applications, namely Solid Oxide Fuel Cells (SOFCs) and Lithium-Ion Batterys (LIBs).

On the basis of the operating point and the impedance of electrochemical systems, a detailed dynamic micro-model will be presented and expanded to a macro-model applicable for SOFCs stacks. This is a precondition for a capable diagnosis system.

The combination of electrochemical characterization techniques, the amendments suggested in this thesis, and the modeling approach shall provide a comprehensive toolbox for the impartial electrochemical characterization of any electrochemical system, whereas the focus is set on SOFCs in this thesis.

Outline

A general introduction into electrochemical systems, with focus on SOFCs and the LIBs as practical examples will be given in chapter 2, followed by a presentation of the techniques and the technology required for their electrochemical characterization in chapter 3. EIS is the central technique this thesis deals with. After a general description of EIS in section 3.5.1, different modifications and amendments are presented. The most important ones are a new algorithm for the derivation of an impedance spectrum out of time domain data in section 3.6.2 and a fast technique to track the change of particular characteristics of the system in the range of seconds or minutes (Single Frequency Electrochemical Impedance Spectroscopy (SFEIS)) in section 3.5.3. The techniques to extract the required system parameters out of the spectra are described in chapter 4. The widely used Complex Nonlinear Least Squares (CNLS) fit introduced in section 4.5.1 is enhanced in this thesis by including the Distribution of Relaxation Times (DRT) (see section 4.4) in the quality criterion, as explained in section 4.5.4. With the results from measurement and parameter identification, the basis of an impedance model for electrochemical systems is established. The setup of such a model is introduced for small-scale or experimental systems and an approach to expand these simple models, in order to enable the simulation of large-scale systems, is introduced in chapter 5. The thesis concludes with a short summary of the obtained results and some remarks of how further improvement of the presented approach can be achieved in chapter 6.

Remarks

The focus of this thesis are electrochemical measurements and their interpretation leading to a system model or to impartial performance data. Therefore explanations of the chemical and thermodynamic background of the system under test, the measurement results, and their interpretation are kept brief or are omitted to allow for deeper discussion of the measurement and identification techniques, but corresponding references will be given.

All measurements in this work were conducted on technically relevant systems. These comprise planar anode supported SOFCs (Anode Supported Cell (ASC)), developed at Forschungszentrum Jülich and commercially available LIBs.

2. Electrochemical Systems

The measurement techniques and modeling approaches presented in this thesis will be explained with the help of practical systems, which are extensively studied at Institut für Werkstoffe der Elektrotechnik (IWE): Lithium-Ion Batterys (LIBs) and Solid Oxide Fuel Cells (SOFCs). These techniques are also applicable to other kinds of electrochemical energy converters, however, a complete introduction for all relevant electrochemical systems seems rather difficult and is the intention of this thesis.

In this chapter, LIBs and SOFCs as representatives for electrochemical energy conversion systems are briefly categorized and explained. For a more extensive introduction to these electrochemical systems, the reader is referred to the corresponding references given. Basic terminology will be introduced as needed to discuss the measurement and modeling approaches that constitute the bulk part of this work[1]. In section 2.1 a general introduction into electrochemical systems will be presented. The fundamental setup and corresponding notations of an electrochemical system are subject to section 2.2. Electrical (section 2.3), thermal (section 2.4) and chemical properties (section 2.5) of electrochemical systems will be introduced, and this chapter concludes with an overview of the modeling approaches relevant for SOFCs and LIBs.

2.1. General Electrochemical Systems

All devices which convert chemical energy into electrical energy and vice versa are part of the scientific discipline of electrochemistry. The term 'system' indicates that the most fundamental practical realization of such an energy converter, often also called a 'cell', can be viewed as a system, that shows a systematic and often quite complex relation between its inputs and its outputs.

The most general definition of an electrochemical system, is that it consists of two electrodes separated by at least one electrolyte phase [1]. The primary difference between an electrochemical reaction occurring in a cell and its complementary redox reaction is that

---

[1]A list of symbols and acronyms is provided in the appendix.

oxidation and reduction do not occur at the same site but separated by the electrolyte as two half-reactions in the so-called 'half-cells' [2].

During a chemical redox reaction one reactant accepts one or more electrons and the other donates one or more electrons. In order to maintain electroneutrality, ions are exchanged. For an electrochemical cell, these ions are exchanged via the ionic conducting electrolyte, that blocks the direct exchange of electrons. These flow through an external conductor [1], where they can do work. The primary application of an electrochemical cell is thus to convert chemical energy into electrical energy.

In many cases these electrochemical reactions are reversible and electric energy can be used to convert a chemical substance with low internal chemical energy into a substance with higher internal chemical energy, like it is done in electrolysis.

A comprehensive and practical introduction into electrochemical systems is given in [2]. The fundamental works [3], [4], and [5] also provide good introductions that reach from the fundamentals of electrochemistry to the components of an electrochemical cell.

If an electrochemical system is able to provide an electrical current, it is called a galvanic cell [5]. There are three types of galvanic cells:

- **Primary galvanic cells.** These are cells that contain the fuel to be converted into electrical energy during operation. This reaction is not reversible. After the fuel is depleted, the cell must be mechanically refilled to provide the possibility of further production of electrical energy. The first galvanic cell invented by Volta in 1800 was of this category [2]. Non-rechargeable batteries or primary batteries also belong to this category [6].

- **Secondary galvanic cells.** The electrochemical reaction associated with this type of cells is reversible. That is why they have two operation modes – discharge, producing electrical energy out of chemical energy, and charge, converting electrical energy to chemical energy. All reactants and products are also contained in the cell. The LIBs discussed in this thesis and, more generally, secondary batteries belong to this category [6].

- **Tertiary galvanic cells.** The fuel for these cells is not contained in the device itself but is supplied externally. Hence, this type of galvanic element is normally referred to as fuel cell. Generally, all types of fuel cells belong to this category and the SOFC, that will be discussed in detail in this thesis, is one of the most prominent one.

If the operation of a tertiary galvanic cell is reversed, it is operated in the so-called electrolysis mode, converting electrical energy into chemical energy and thereby producing fuel. This fuel may be externally stored and utilized later in the same cell operated in fuel cell mode to produce electrical energy, but is not restricted to this.

2.1.1. Solid Oxide Fuel Cell (SOFC)

SOFCs are fuels cell which utilize a solid ceramic oxide electrolyte. At elevated temperature, typically $T = 600 \ldots 1000\,°C$, acceptable ionic conductivity is achieved. The requisite high temperatures for these devices, plus the catalytic activity of the anode, enable the use of carbon monoxide, methane and other hydrocarbons or diesel reformat as fuel gas. Operating temperatures range from $T = 600 \ldots 1000\,°C$.

2.1.2. Lithium-Ion Battery (LIB)

LIBs use Lithium intercalation materials. The electronegativities for these materials cover an extremely large range and consequently, it is possible to assemble a battery with a high Open Circuit Voltage (OCV) by using a material with a very low electronegativity for the anode and a material with a very high electronegativity for the cathode. Moreover, Lithium is a very light material and therefore LIBs are capable of reaching a very high weight-specific energy density, an important feature for mobile applications.

The use of different possible liquid or solid electrolytes determines the possible operating temperatures, that are the most narrowed for liquid electrolytes at $T = -10 \ldots 60\,°C$. Solid electrolytes can be operated at lower temperatures in principle. However, the ionic conductivity decreases dramatically with decreasing temperature.

2.2. Setup of Electrochemical Systems

An electrochemical system consists of three basic components: the electrolyte, anode and cathode. In section 2.2.1 the major tasks for the three components are explained for all electrochemical cells that can be categorized as batteries or fuel cells. Also other important components are required for a practical realization of an electrochemical system are described there. Additional information regarding the system setup and denotations applied for the components is provided in section 2.2.2. The electrochemical reaction in an electrochemical systems follows a fundamental law, the conservation of species, that is introduced in section 2.2.3.

2.2.1. Components of Electrochemical Systems

The necessary components of an electrochemical system are electrolyte, anode, and cathode. Their description is kept general in sections 2.2.1.1 to 2.2.1.3. A more specific description of the materials used for LIBs and SOFCs can be found in section 2.5.1. As additional components for the practical realization of an electrochemical system, current collectors and the housing are briefly addressed in section 2.2.1.4.

2.2.1.1. Electrolyte

The electrolyte is the defining part in an electrochemical system. Its primary purpose is to separate the anode and cathode volume spatially to prevent a spontaneous reaction of the reactants. But by its characteristics it defines the charge carriers and the operating temperature and as a consequence also the fuels (fuel cells) and electrode materials.

The electrolyte must be an electronic insulator[2] to prevent an internal short circuit. Even small electronic conductivities are unfavorable, because the energy of the current through

---

[2]Electronic insulation means negligible electronic conductivity. Electronic conductivity is a measure for the ability to transport electrons as opposed to ionic and proton conductivity, which describe the ability to transport ions and protons, respectively. Electrolyte materials require high transference numbers for ionic conductivity [7].

the electrolyte cannot be used in the external circuit and is lost as heat to the surrounding environment.

The term 'volume' is used here with intent because the reactants and products in the electrodes can be gaseous (e. g. in a SOFC), solid (e. g. in a LIB) or even liquid (e. g. in a Direct Methanol Fuel Cell (DMFC)). The electrolyte itself can either take the form of a solid (e. g. in a SOFC) or liquid (e. g. in a LIB).

The second purpose of the electrolyte is to enable the direct transport of ions or charged molecules, while the residual electrons and holes provide electrical energy for the external electric circuit.

Materials, charge, and charged species that meet these requirements vary and this explains the many types of galvanic cells. A comprehensive list of different types of cells and the corresponding electrolyte materials, charge, and charged species can be found in [8] for batteries and in [9] for fuel cells.

2.2.1.2. Anode

The anodic reaction is an oxidation providing free electrons on the anode side. During operation the anode has a negative charge and acts as the negative pole in the electrochemical system as shown in figure 2.1.

The anode must be electronically conductive to provide the electrical connection to the external electrical circuit. The second requirement is that the anode is responsible for the charge transfer from ions to electrons (for electrolytes conductive for negatively charged ions or molecules) or vice versa (for electrolytes conductive for positively charged ions, molecules or protons).

For some configurations the charge carriers are incorporated in the anode material (as Li ions are in LIB anodes), for others the anode has to be catalytically active to facilitate the reaction and to lower the polarization losses (as the Ni in SOFC anodes).

2.2.1.3. Cathode

In principle the cathode has the same function and requirements as the anode, only that the chemical reaction is a reduction and electrons are bound by the cathodic half reaction, so that holes and therefore a positive charge and positive pole emerges during the electrochemical reaction. It has to be electronically conductive and conduct the charge transfer from electrons to ions (for electrolytes conductive for negatively charged ions or molecules) or vice versa (for electrolytes conductive for positively charged ions, molecules or protons).

A further benefit of a good cathode is to facilitate the cathodic reaction by catalytic activity. The requirements being the same as for the anode, the materials and microstructural properties that meet these requirements best will differ.

2.2.1.4. Additional Components

In order to make an electrochemical system applicable, two other important components have to be added:

- **Current Collectors.** The role of a current collector is to provide the electronic connection of the electrodes to the outer circuit. The requirements for a current collector is high electronic conductivity as well as chemical compatibility with the materials and the ambient conditions of the cell components they are connected with. Due to the oxidizing environment on the cathode side and the reducing environment on the anode side, it is common to use different materials as current collector, respectively. For example, Cu is used to contact the anode of LIBs and Al for contacting the cathode.

  For SOFCs the term current collector is commonly also used for an additional highly porous electrode layer, that shows a good electronic conductivity. Such a layer is used for nano-sized cathodes with a thickness of only a few hundred nanometers, for example. These usually show lower porosities than standard electrodes and a direct metallic contacting is not easily achieved [10]. That is why an additional current collector layer is applied between electrode and metallic current collector.

- **Housing.** As described in section 2.1, in every electrochemical system an electrochemical reaction is split into two half cell reactions by the electrolyte. In order to avoid direct contact between the two reactants, the electrolyte has to be both impermeable and inert for these substances, but also a housing has to be provided, that prevents direct reactions among the reactants and with the surrounding environment. As the operating parameters for LIBs and SOFC are very different, also the requirements for the housings differ. That is why they are treated separately here.

  SOFC/SOEC. High operating temperatures and gaseous reactants require the housing for a commercial SOFCs be thermally insulating and impermeable. A description of a very advanced stack design by Forschungszentrum Jülich (FZJ) is presented in [11]. The housing used for research on single cells is described in section 3.1.2.

  LIB. For the already widely commercially available LIB, three different types of housings exist [12]: cylindrical cells, prismatic cells, and pouch cells. In laboratory scale tests, also one-way or reusable metal housings are used. They are mentioned in section 5.1.1.

2.2.2. System Setup and Notations

The general definitions and purposes of electrolyte, anode and cathode have been introduced in section 2.2.1. In practice these definitions are not satisfactory, because most electrochemical systems are also used in reversed operation. That is why this section is needed to provide unambiguous denominations for all electrodes dealt with in this thesis.

2.2.2.1. SOFC

Their first application as fuel cell has set the notation of anode and cathode that will be used in this thesis for SOFCs. So the definition from section 2.2.1 holds in this case and the negative electronic pole is called anode, and the positive electronic pole is the

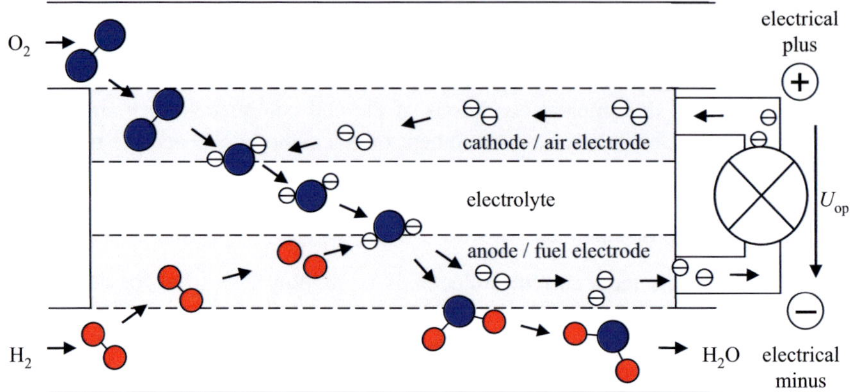

Figure 2.1.: Function principle and notations of a SOFC. The operating voltage is $U_{op} = U_{OCV} - \sum_k \eta_k$, with $\eta_k > 0$ as introduced in section 2.3.2.

cathode, as shown in figure 2.1. The denomination is also strongly connected to the applied materials for both electrodes. These materials are often referred to as anode or cathode materials.

However, it should be noted that the emergence of an operation of SOFCs in electrolyzer mode led to a more general denomination of fuel electrode for the anode and oxygen electrode for the cathode [13]. This notation holds for both operation modes, because fuel and oxidant gas are supplied on the same electrodes in both modes, respectively, but nevertheless, in SOFC research the two electrodes are traditionally called anode and cathode.

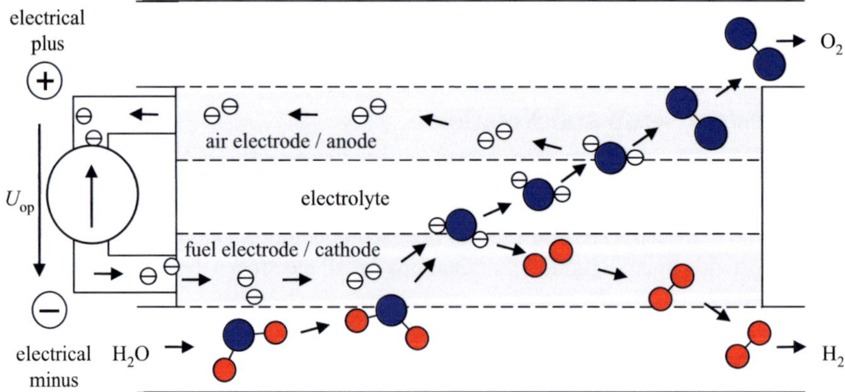

Figure 2.2.: Function principle and notations of a SOEC. The operating voltage is $U_{op} = U_{OCV} - \sum_k \eta_k$, with $\eta_k < 0$ as introduced in section 2.3.2.

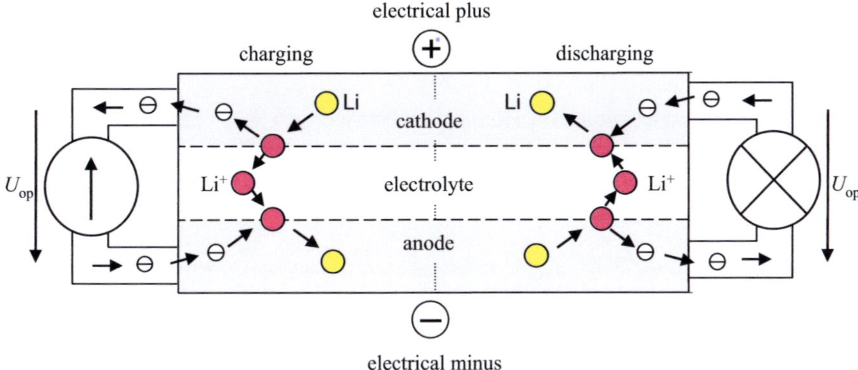

Figure 2.3.: Function principle of a LIB. The operating voltage is $U_{\mathrm{op}} = U_{\mathrm{OCV}} - \sum_k \eta_k$, with $\eta_k < 0$ for charging and $\eta_k > 0$ for discharging as introduced in section 2.3.2.

2.2.2.2. SOEC

In Solid Oxide Electrolysis Cell (SOEC) mode anode and cathode are switch by the general definition in section 2.2.1, as shown in figure 2.2. However, this would be confusing because the notation of anode and cathode for SOFC mode has a much longer history. But strictly speaking, this notation is wrong for SOEC systems. As a compromise, it is common to speak of fuel electrode and oxygen electrode in SOEC systems. This holds for both SOFC and SOEC mode, as explained in the previous section. And even though it has not become widely accepted in SOFC research, SOEC research in general and also this thesis use the terms fuel and oxygen electrode for SOECs.

2.2.2.3. LIB

LIBs have been designed for two operation modes - charge and discharge. When charging a LIB, the oxidation reaction takes place at the opposite electrode, as it does during discharging, as shown in figure 2.3. Consequently, the denominations anode and cathode should switch, when the operation mode changes. Nevertheless, it is more practical to have a permanent notation for both electrodes. By convention, the discharge mode sets the denomination, as indicated in figure 2.3. This also means, that the negative electronic pole is always called anode, and the positive electronic pole is the cathode.

Even though the external current and the internal flow direction of the Li ions change for both operation modes, the electrical potential difference, the cell voltage has always the same sign. The characteristics of this voltage and how it is created is described in section 2.3.

2.2.3. Conservation of Species

Not only is it important for an electrochemical system, in which direction and to what amount the different species – ions, protons and electrons – flow. The electrochemical reaction is also correlated to a conversion of chemical compounds. The amount of a chemical substance is commonly given as its volume V or mass m. This amount can be converted to an electrical charge Q by the electrochemical system.

The mathematical conversion of a chemical substance into an electrical charge can be done by using the conservation of species. It states, that the charge exchanged between anode and cathode through the electrolyte has to correspond to the number of molecules or atoms that have reacted on anode and cathode side.

In H_2/H_2O operation for SOFCs, this means that one oxygen ion with a double negative charge, that passes through the electrolyte, has to emerge from the half of an oxygen molecule on the cathode side and result in one water molecule formed on the anode side. This way, a current drawn can be correlated to a gas volume utilized. In turn, a volume of fuel gas consumed can be correlated to an equivalent current.

For LIBs the amount of reaction partners in the anode and cathode electrode define the capacity of the battery, namely the charge Q, that can be drawn during one cycle. If there is a noticeable residual electronic conductivity present in the electrolyte, not all of the charge created by the electrochemical reaction can be used in the outer circuit. The ratio between usable charge and created charge is called faradaic efficiency η_F in [14] for SOFCs. It is hard to measure but can be assumed to $\eta_F = 1$, because of the very low residual electronic conductivity of the standard electrolyte materials used for SOFCs. For LIBs it can be measured directly, because it is the ratio between the charge, that has flown during discharging the cell, $Q_{\text{discharge}}$, and the charge, that has flown during charging the cell, Q_{charge}, and is often called coulombic efficiency η_Q and its value is commonly very high, too:

$$\eta_Q = \frac{Q_{\text{discharge}}}{Q_{\text{charge}}} = 0.95\ldots 1. \tag{2.1}$$

In general, for $\eta_F = 1$ or $\eta_Q = 1$, the calculation of the electrical charge Q provided by the conversion of the mass of a chemical substance m usually used for LIBs is given by:

$$Q = \frac{m}{M} \cdot z \cdot F \quad \text{with} \quad F = e \cdot N_A. \tag{2.2}$$

The charge Q with respect to the gas flow Q_{gas} (volume V per time) consumed is often considered for SOFCs and is given by:

$$Q = \frac{Q_{\text{gas}}}{V_m} \cdot z \cdot F. \tag{2.3}$$

Due to the different number of electrons per converted atom or molecule the charge number z can have different values for different half cell reactions. All half cell reactions relevant for this thesis are listed together with the corresponding ions and the resulting value for z in table 2.1.

For tertiary galvanic cells, reactants and products have to be supplied to and removed from the reaction sites constantly. As this mass flow can always be converted into an electrical charge per time, this mass flow is also referred to as the equivalent current I_{eq}.

system	electrode	half cell reaction	z
LIB	both	$Li \rightleftharpoons Li^+ + e^-$	1
SOFC/SOEC	fuel electrode	$H_2 + O^{2-} \rightleftharpoons H_2O + 2e^-$	2
SOFC/SOEC	air electrode	$O^{2-} \rightleftharpoons \frac{1}{2}O_2 + 2e^-$	2

Table 2.1.: Half cell reactions of the electrochemical systems discussed in this thesis together with the corresponding ions and the resulting value for the charge number z.

2.3. Electrical Properties of Electrochemical Systems

Like other sources of electrical energy, electrochemical systems are characterized by the terminal voltage U_{op}. The origin of U_{op} will be explained in section 2.3.1. The static behavior in the presence of a constant current density J^3 is content of section 2.3.2. The dynamic relation of current and voltage is given by the complex and frequency dependent impedance $Z(\omega)$ that will be explained in section 2.3.3.

2.3.1. Open Circuit Voltage (OCV)

In this section the thermodynamic derivation of the OCV will be given in brief. It is based on [3]. Other comprehensive derivations can be found in [2, 14]. A deeper insight into the chemical context can be found in [15]. In this thesis, no distinction is drawn between the theoretical cell voltage and the terminal voltage under open circuit conditions U_{OCV}, as will be explained in section 4.1.1.

In a galvanic cell, chemical energy is converted into electrical energy. This means there is a general chemical reaction of the type

$$\underbrace{A + B}_{I} \rightarrow \underbrace{C + D}_{II} \tag{2.4}$$

involved, associated with a change in free molar reaction enthalpy. ΔG is the corresponding difference in molar free energy or Gibbs free energy. It is thus the maximum electrical energy that can be obtained from the transformation of one mole of reactants during a chemical reaction. At least the same amount of energy has to be introduced into the system for the reverse reaction, if applicable. ΔG^0 is the standard value for ΔG. A list of values for ΔG^0 relevant for different fuel cell reactions can be found in [14, 16]. For LIBs the standard potentials of the applied materials are listed in [6, 17]. From these values, the standard cell voltage U_{OCV}^0 can be calculated, as will be shown in equations (2.9) to (2.11). For a deeper insight into the energies associated with the electrochemical reaction, see [2, 3, 14].

---

[3]In this thesis both physical quantities current and current density will be used. As it is common for electrochemical systems, the current density J will be used to describe the movement of charged species within the electrochemical system. For the measurement techniques described in chapter 3 and for the model derived in section 5 the current I is used for any current outside the electrochemical system. Dynamic current densities and currents are indicated by lower-case variables, respectively.

In order to understand the correlation of the electrical cell voltage U_{OCV} or OCV not assuming standard conditions and the chemical potential, the electrochemical potential $\bar{\mu}$ has to be introduced,

$$\bar{\mu}_n = \mu_n + z_n F \varphi, \tag{2.5}$$

where μ_n is the chemical potential of the component n [4],

$$\mu_n = \mu_n^0 + RT \ln a_n. \tag{2.6}$$

The chemical potential consists of the standard value μ_n^0 at standard conditions and a term, that depends on the concentration of component n. The latter part is determined by the chemical activity a, which is a measure of the effective concentration [4]. While electrochemical equilibrium is assumed,

$$\bar{\mu}_n(\mathrm{I}) = \bar{\mu}_n(\mathrm{II}), \tag{2.7}$$

the electrical potential difference $\Delta\varphi$ or cell voltage U_{OCV} can be calculated by the so-called Nernst equation

$$U_{\mathrm{OCV}} = \Delta\varphi = \underbrace{\frac{\mu_{n,\mathrm{I}}^0 - \mu_{n,\mathrm{II}}^0}{zF}}_{\text{material specific}} + \underbrace{\frac{RT}{zF} \ln \frac{\prod_n a_{n,\mathrm{I}}}{\prod_n a_{n,\mathrm{II}}}}_{\text{concentration specific}}. \tag{2.8}$$

The material specific part of equation (2.8) equals

$$\frac{\mu_{n,\mathrm{I}}^0 - \mu_{n,\mathrm{II}}^0}{zF} = \frac{\Delta G^0}{zF}, \tag{2.9}$$

because the sum of changed chemical potentials through the chemical reaction weighted by the stoichiometric factor ν equals the change in Gibbs free energy:

$$\Delta G = \sum_n \nu_n \mu_n. \tag{2.10}$$

Therefore the cell voltage can be linked directly to the change in reaction enthalpy:

$$U_{\mathrm{OCV}} = -\frac{\Delta G}{zF}, \quad \text{or for standard conditions} \quad U_{\mathrm{OCV}}^0 = -\frac{\Delta G^0}{zF}. \tag{2.11}$$

SOFC

In a SOFC, the reactants and products are present as gases, that are continuously supplied to and removed from the electrodes. Therefore a change in U_{OCV} of a SOFC is primarily dependent on the concentration specific part of equation (2.8).

A simplification of equation (2.8) can be introduced for gaseous redox partners, because in this case, the chemical activity a is proportional to the oxygen partial pressure p_{O_2}. If the chemical potentials of each half reaction are formulated and inserted in equation (2.8) as demonstrated in [14, 18], this yields a simplified version of the Nernst equation:

$$U_{\mathrm{OCV}} = \frac{RT}{zF} \ln \frac{\sqrt{p_{\mathrm{O}_2,\text{cathode}}}}{\sqrt{p_{\mathrm{O}_2,\text{anode}}}}. \tag{2.12}$$

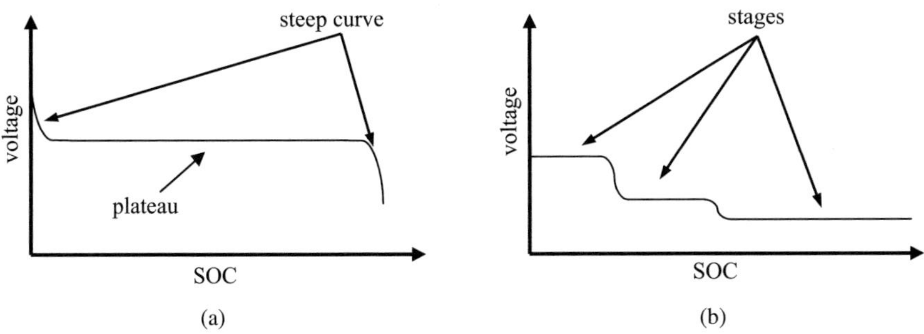

Figure 2.4.: Schemes of characteristic OCV curves for LIBs showing (a) steep parts and a plateau and (b) different stages.

It is further shown in [14, 18], that equation (2.8) can also be rewritten as

$$U_{\text{OCV}} = -\frac{\Delta G^0}{zF} + \frac{RT}{zF} \ln \left(\frac{\sqrt{p_{\text{O}_2,\text{cathode}}} \cdot p_{\text{H}_2,\text{anode}}}{p_{\text{H}_2\text{O},\text{anode}}} \right). \tag{2.13}$$

Equation (2.13) has the advantage that the cell voltage U_{OCV} can be calculated as function of the supplied gas concentrations at anode and cathode side for $\text{H}_2/\text{H}_2\text{O}$ operation.

LIB

A comprehensive explanation of the OCV of LIBs for all applicable material configurations is quite a challenging task. It can be explained comprehensively by the introduction of phase diagrams as done in [7], but will be omitted here.

In general, the OCV of LIBs is determined by the standard potentials of the materials applied [6]. But it changes with the State of Charge (SOC) in a complex manner. However, a brief summary of observable shapes of the OCV curve $U_{\text{OCV}}(SOC)$ together with the corresponding explanations is provided here with the help of equation (2.8).

- **Steep parts.** Towards $SOC = 0$ and $SOC = 1$, it is characteristic for the OCV curve to show steep gradients, as illustrated in figure 2.4(a). That is where the concentration of Li in the electrodes is near its minimum or maximum. As the concentration of Li at the surface of the electrodes and at the interface to the electrolyte increases or decreases significantly, the change in U_{OCV} is mainly governed by the concentration specific part of equation (2.8). This behavior can be observed for all electrode materials.

- **Plateaus.** Large parts of the OCV curve of LIBs often show a very flat behavior, as also indicated in figure 2.4(a). This is the case when a the incorporation of Li into the electrode is associated with a phase transition. Then the concentration of Li at the surface of the electrode does not change significantly. More significantly, the active material in the electrode is continuously converted and the interface of the two phases migrates through the electrode material. On its surface the same potential is present for as long as both phases are present in the electrode. In this case, U_{OCV} is mainly governed by the material specific part of equation (2.8). A very pronounced plateau is characteristic for the OCV curve of the cathode material $LiFePO_4$.

- **Stages.** If more than one phase transition is possible in an electrode material, the U_{OCV} curve will show multiple stages punctuated by steep transitions, as indicated in figure 2.4(b). The most prominent example for such a behavior is the graphite anode [7]. Within the different stages the material specific part of equation (2.8) determines the OCV and at the transitions between the stages, the changing concentration of Li is responsible for the change in OCV through the concentration specific part of equation (2.8).

2.3.2. Cell Voltage under Load

In a first approximation, the cell voltage of an electrochemical system behaves like a real voltage source – an ideal voltage source in series with a resistor, the so-called inner resistance of the voltage source. If a current is drawn from the real voltage source, the terminal voltage decreases. The reason for this behavior is the voltage drop at the inner resistance.

In electrochemical systems a similar voltage drop is observable, which is due to ohmic and polarization losses. The voltage drop at the electrodes is called overpotential η in electrochemistry [3][4], and $\eta_{tot} = U_{OCV} - U_{op}$ is defined as the total overpotential in this thesis. It is caused by the electrochemical reaction itself, which can be described by Butler-Vollmer kinetics or by the Tafel equation [1], or by transport mechanism such as solid state or gas diffusion. A comprehensive overview of the polarization processes relevant for SOFCs is provided in [18]. For LIBs this is more difficult due to the different material systems and configurations.

For small currents the voltage decrease is approximately linear and the Current/Voltage (C/V) characteristics are given as a decreasing line intersecting the ordinate at U_{OCV}.

Even though the equations describing the OCV and the voltage under load are valid for both SOFCs and LIBs, the static behavior under load is different for both systems. It is not possible to adjust a stable operating point with a current density $|J| > 0$ for a battery because of the associated change in SOC. Therefore the static C/V diagram is not a well-established display for the performance analysis for LIBs. Hence, the C/V characteristics are introduced as a fundamental characteristic diagram for fuel cells in section 2.3.2.1. The same relations account for batteries, but are more difficult to determine and to handle. This will be explained in section 2.3.2.2. Charge/discharge characteristics are often addressed as substitute for the C/V characteristics of batteries. This will be discussed in section 2.3.2.3. The temperature effects that are related to a current density through the system are subject to section 2.4.2.

2.3.2.1. C/V Characteristics of Fuel Cells

The nonlinear behavior exhibited in C/V curves necessitates the definition of several types of resistances to appropriately characterize the cell behavior. Both definitions and the use

---

[4]For reasons of clarity, the voltage drop at the electrolyte is defined as η_Ω. In [3] this voltage drop is called iR-drop.

of these types are not consistent in the literature. The following list introduces the three most important parameters, as they are treated in this thesis.

- *ASR.* The Area Specific Resistance (ASR) is a common measure to judge the performance of an electrochemical system. It is the ohmic resistance normalized to a square-centimeter, so that the unit for the ASR is Ωcm^2:

$$ASR = \frac{\Delta U}{\Delta J}. \tag{2.14}$$

A well-performing electrochemical system is characterized by a small ASR, i. e. the cell voltage decreases only slightly, if a current density is adjusted. Values for the ASR are often found in the literature. In this thesis the ASR is defined as the impedance for $\omega \to 0$ or Z_DC at OCV. This is in agreement with the major part of publications on SOFCs.

- Z_DC. Generally, the Direct Current (DC) impedance Z_DC is defined as the impedance in an operation point for very low frequencies (see also figure 2.15(a)). Because of the nonlinearities in the C/V curve of fuel cells, Z_DC is a function of the current density J as indicated in figure 2.5. For $J = 0$, the DC impedance equals the ASR, $Z_\mathrm{DC}(0) = ASR$. In this thesis, no distinction is made between impedance and area specific impedance. Both are described by the same symbol Z.

- R_i. The inner resistance R_i is defined as

$$R_\mathrm{i} = \frac{U_\mathrm{OCV} - U_\mathrm{op}}{J_\mathrm{op}}. \tag{2.15}$$

The term 'area specific' is omitted here for two reasons: to avoid confusion with the ASR, for which a clear definition was given above, and because it is more commonly used for the analysis of a system behavior rather than for electrochemical analysis. However, the inner resistance R_i is an appropriate criterion for the performance of an electrochemical system, as it represents the losses in the operating point[5]. However, it strongly depends on the operation point as indicated in

---

[5]Operating point in this thesis is determined by the operating conditions (temperature, SOC, gases) plus either current density or voltage adjusted.

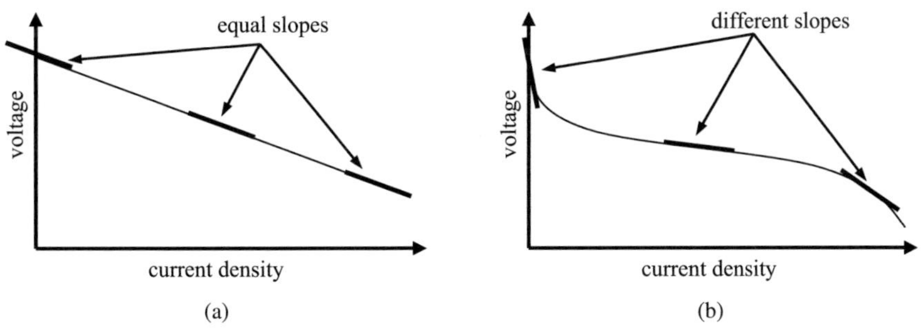

Figure 2.5.: Schemes of (a) a linear and (b) a nonlinear C/V curve with the corresponding slopes indicated by the thick lines.

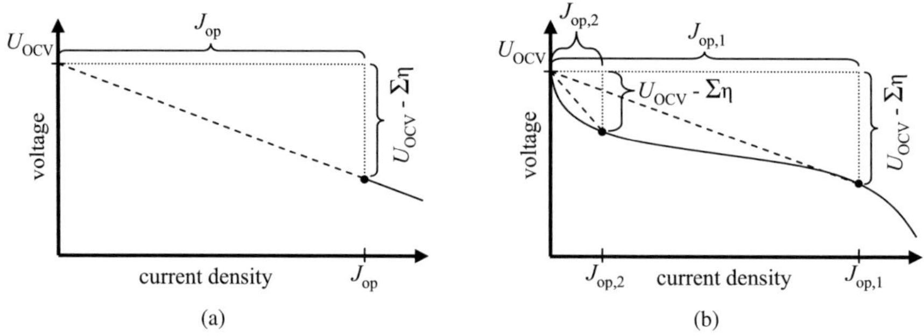

Figure 2.6.: Schemes of (a) a linear and (b) a nonlinear C/V curve indicating different inner resistances obtained at different operating points for nonlinear C/V curves. The inner resistance R_i is given by the slope of the dashed lines, respectively.

figure 2.6. Therefore it is necessary to specify an operation point for every R_i given. This can either be the current density or the operating voltage.

Finally, three remarks concerning the C/V characteristics of fuel cells should be mentioned here:

- **Overpotential.** The total overpotential η_{tot} consists of different losses, that will be explained for SOFCs in section 5.3.2:

$$\eta_{\text{tot}} = \sum_k \eta_k. \tag{2.16}$$

  Every individual overpotential $\eta_k(J)$ corresponds to a contribution to $Z_{\text{DC}}(J)$: $Z_{\text{DC},k}(J)$. As a nonlinear system has to be assumed, $\eta_k(J) \neq J \cdot Z_{\text{DC},k}(J)$, as will also be explained in section 5.3.2. The overpotential is positive for SOFC operation and LIB discharge and negative for SOEC operation and LIB charge [1].

- **Operating conditions.** If two electrochemical systems shall be compared with the help of the C/V characteristics, it is essential to compare the above introduced values for identical operating conditions.

- **SOFC and SOEC mode.** Even though the current density J has a negative sign for SOEC mode, all resistance values used in this thesis have a positive sign, because also $\Delta U = U_{\text{OCV}} - U_{\text{op}}$ has a negative sign in SOEC mode and LIB charge mode.

2.3.2.2. C/V Characteristics of Batteries

As pointed out before, C/V characteristics are also relevant for batteries and can be used to measure their performance. However, there are two problems concerning the measurement, the determination, and the illustration of the C/V characteristics for batteries.

First, the C/V characteristics differ for every SOC. This means, that the C/V characteristics are represented by a set of curves. Up to date, there is no widely accepted convention and

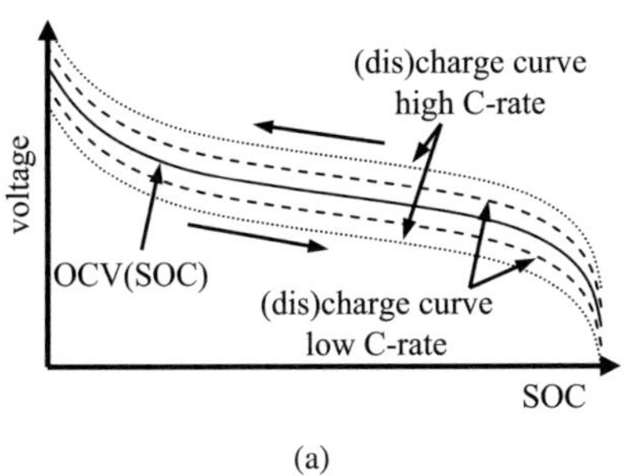

(a)

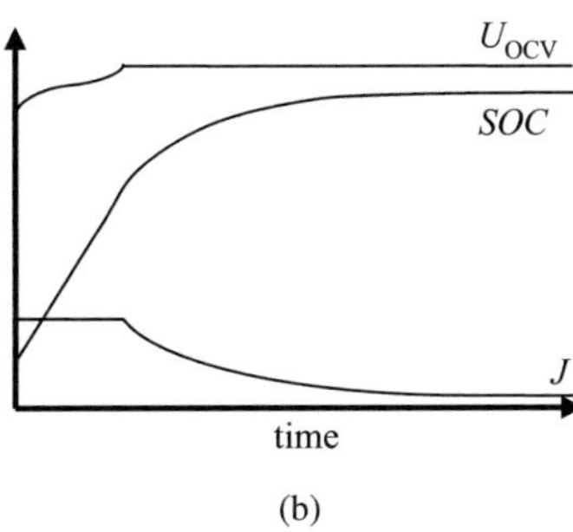

(b)

Figure 2.7.: (a) Schemes of charge and discharge curves with varying C rate. (b) Scheme of a CCCV charge procedure.

not even a discussion, for which SOC a characteristic C/V curve should be measured and presented in order to facilitate a comparison of different batteries.

The second problem is more significant and is due to the following conflict:

- Each measurement point for the C/V characteristics must be determined in steady state conditions. To realize this, a constant operating point at a given current density has to be adjusted for a certain period of time, until the dynamics have decayed and a constant cell voltage is obtained. The measurement is more accurate the longer the settling time before the voltage value is recorded.

- An applied constant current will change the SOC of the battery and results in a more or less pronounced change in cell voltage. The measurement is more accurate the shorter the settling time before the voltage value is recorded.

These difficulties are the reason why the battery performance is more commonly evaluated with the help of their charge/discharge characteristics, which are subject to the next section.

2.3.2.3. Charge/Discharge Characteristics of Batteries

The charge/discharge characteristics of a battery simply consist of the charge curve and the discharge curve. The performance of a battery can be rated by those curves. They exhibit the course of the cell voltage U_{op} during a charge or discharge procedure maintaining a constant current density or C rate[6]. While the OCV is determined by the material properties of the battery, performance can be determined by the deviation of the operating voltage U_{op} from the OCV at a constant C rate. For every C rate, a different charge/discharge curve is obtained as shown in figure 2.7(a). This is explained in great detail in [19]. A practical example can be found in [20]. In principle, important cell parameters like the DC impedance Z_{DC} can be deduced from these curves. In practice, this is rarely done.

---

[6]The C rate is a special convention for charge and discharge procedures for batteries, and is a value normalized to the capacity of the battery being tested. A C rate of $1C$ means, that the current or current density is chosen so that the battery is charged or discharged within one hour. Accordingly, with a charge C rate of $5C$ the cell is charged within $1/5$ of an hour and with a discharge rate of $C/5$, the cell is discharged in 5 hours.

Another common technique that should be mentioned in this context is known as Constant Current Constant Voltage (CCCV). For charging the battery, a constant current is applied, until the end-of-charge voltage is reached. After that, this voltage is kept constant, until the current has dropped below a certain value, as depicted in figure 2.7(b). The corresponding discharge is conducted with a constant current until the end-of-discharge voltage is reached. However, this technique is normally not used for the characterization of the battery but only to perform defined charge/discharge cycles.

2.3.3. Impedance of Electrochemical Systems

As mentioned above, the cell voltage U_{op} decreases due to ohmic and polarization losses, if a positive current density is adjusted. More generally, in every electrochemical system, electrochemical processes occur, that are responsible for a voltage drop during operation and constitute a loss process. In the static characteristics introduced in section 2.3.2, these processes are only represented as a sum and cannot be separated.

Each of these losses are characterized by their relaxation time, or characteristic frequency, and their ASR $Z_{\mathrm{DC}}(J)$. In other words, there are fast and slow processes; they have different rates. The impedance response can be reproduced by basic electrical components, that constitute a so-called Equivalent Circuit Model (ECM).

The literature on this topic is extremely extensive and versatile. The most cited, and probably most fundamental, work is [21]. A more recent and more practical textbook is [22]. In [23], a rather brief but still comprehensive introduction into the topic is given. Apart from the information given in these publications, a large number of review and overview articles will be recommended in the corresponding subsections, as they become relevant to the discussion.

Although the literature is extensive concerning the mathematical description and the underlying differential equations, practical explanations for the different time constants that arise in an electrochemical system, remain scarce. Here, a short and comprehensive explanation of how the complex impedance of an electrochemical system originates, what basic steps are responsible, and why current and voltage to behave as they do will be attempted.

In the first part of this section, the definition of the electrochemical impedance is provided. Basic impedance elements are presented in section 2.3.3.2. This is followed by an attempt to explain impedance more practically in section 2.3.3.3. Then, the impedance elements relevant for the systems in this thesis will be introduced in sections 2.3.3.4 and 2.3.3.5. This section is completed by some concluding remarks in section 2.3.3.6.

2.3.3.1. Definition of the Electrochemical Impedance

The electrical impedance, Z, of a system is defined as its complex and frequency dependent resistance, the correlation between voltage and current. This is also true for the electrochemical impedance (also named Z) with the difference that the electrochemical impedance is commonly normalized and defines the correlation between voltage and current density. The corresponding physical unit is $\Omega\mathrm{cm}^2$.

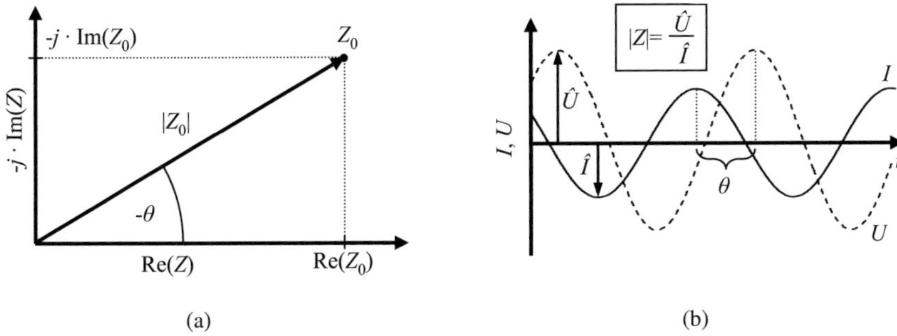

Figure 2.8.: Schemes of (a) the complex quantity Z_0 and (b) sinusoidal voltage perturbation and the corresponding current response indicating the meaning of the absolute value of the impedance $|Z_0|$ and the phase angle θ.

It is complex because of the dynamics of the electrochemical system. This means that the current response of a voltage excitation[7] is not a proportionate (does not follow Ohm's law) of the excitation, but also depends on the previous values. For a given frequency the complex impedance is given by an absolute value, $|Z(\omega)|$, which accounts for the ratio between the amplitudes of the excitation and response signals, and the phase angle $\theta(\omega)$ gives a measure for the lag between those two signals (see figure 2.8). This can be explained best by a harmonic oscillation with an angular frequency, ω. For the purpose of analyzing electrochemical systems it is more convenient to work with real, $\mathrm{Re}(Z)$, and imaginary, $\mathrm{Im}(Z)$, parts of the impedance. These values can be easily derived from the so-called polar form given by $|Z(\omega)|$ and $\theta(\omega)$:

$$\mathrm{Re}(Z(\omega)) = |Z(\omega)|\cos(\theta(\omega)) \quad \text{and} \quad \mathrm{Im}(Z(\omega)) = |Z(\omega)|\sin(\theta(\omega)). \qquad (2.17)$$

There are several display types used to visualize the impedance, which will be explained in detail in section 4.2. In this thesis, Nyquist plots will be primarily used. In these, the negative imaginary part of the impedance is plotted against its real part.

2.3.3.2. Basic Impedance Elements

The three basic impedance elements resistor, capacitor and inductor will be introduced in this section (see also figure 2.9). Additionally, the most fundamental connection for impedance analysis – the RC circuit, is presented here. A detailed explanation of these components can be found in [24]. In section 2.3.3.5 more complex impedance elements are discussed, which are more specific to electrochemical systems.

Resistor

An ohmic resistor is the most simple element, because its impedance is constant for all frequencies and has no complex part:

$$Z(\omega) = R. \qquad (2.18)$$

---

[7]Or vice versa the voltage response of a current excitation.

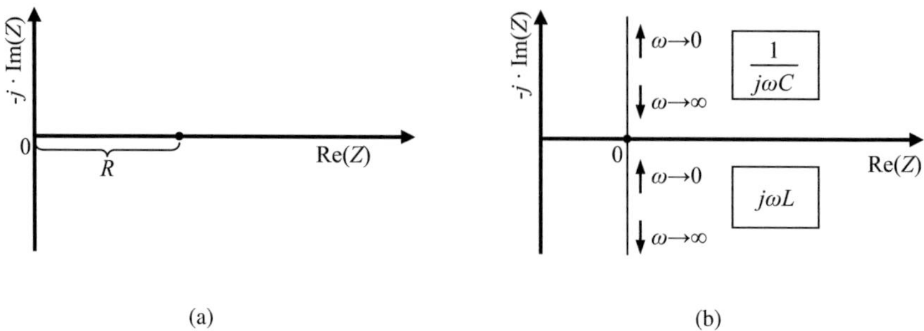

Figure 2.9.: Schemes of the impedance Z of (a) a resistor and (b) of a capacitor and an inductor shown as a Nyquist plot.

Capacitor

The impedance of an ideal capacitor has no real part and is therefore loss-free. It is defined as a negative complex number, that is converging to zero for high frequencies:

$$Z(\omega) = \frac{1}{j\omega C}. \tag{2.19}$$

Inductor

An ideal inductor is also loss free and shows an impedance, that is purely complex and strictly positive. For $\omega = 0$ it is zero and increases for higher frequencies:

$$Z(\omega) = j\omega L. \tag{2.20}$$

An inductivity in the impedance spectrum usually originates from external sources, such as the inductive behavior of the wiring providing the electrical connection of the measurement device with the electrochemical cell. It is sensible to model this artifact with an inductor and to correct the measured spectrum by subtracting its impedance contribution subsequently from the impedance curve. The error from suboptimal wiring is removed by this, if it can be modeled by a pure inductivity. Unfortunately, measurement results commonly show a much more complex behavior for high frequencies, which requires more complex models for the inductivity, as discussed in [25].

RC Circuit

An RC circuit is a parallel connection of a resistor and a capacitor (see figure 2.10(a)) and is a fundamental element, on which basis most impedance responses of electrochemical systems can be reconstructed[8]. Its complex impedance is given as (see also figure 2.10(b)):

$$Z_{RC}(\omega) = \frac{R_{RC}}{1 + j\omega\tau_{RC}}, \quad \text{with} \quad \tau_{RC} = \frac{1}{\omega_{RC}} = R_{RC}C_{RC}. \tag{2.21}$$

---

[8]In fact, most technical process can be approximated by this impedance element. In control engineering, the corresponding behavior is also called PT_1 characteristics [26].

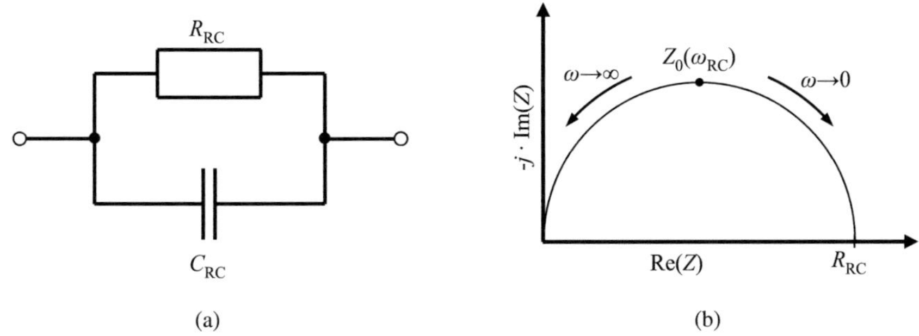

Figure 2.10.: (a) Symbol of an RC circuit and (b) Nyquist plot of an RC circuit.

The time constant τ_{RC} is the inverse of the angular frequency ω[9]. A more detailed explanation of the impedance response of an RC circuit will be given in following section 2.3.3.3.

2.3.3.3. Practical Explanation of the Impedance of a Basic Electrochemical System

All passive electrochemical processes can be broken down to a (not necessarily finite) combination of RC circuits[10], whose response is given by equation (2.21). In this section, the impedance response of an electrochemical system is compared to the simplified response of an RC circuit. These responses take the form of the Nyquist plot shown in figure 2.10(b). The behavior of the real and the imaginary part of the impedance shown in figure 2.11 is discussed as it relates to electrochemical systems.

Very high frequencies. At very high frequencies only little charge is transported during one half cycle of a sinusoidal perturbation. Since this charge can be provided by the capacitor without any losses, the RC circuit behaves like a short-circuit. In an electrochemical system, charge is only slightly dislocated. The time of one half cycle is far too small to activate a loss process, hence, the impedance is zero, $Z(\omega \rightarrow \infty) = 0$.

High frequencies. For slightly lower frequencies, the capacitive behavior is visible, because the complex part decreases rapidly from zero to a negative value. The angle with the real axis at the high frequency intersection is $90°$. The real part of the impedance also increases, but this increase proceeds slowly, because the capacitor can still accept the majority of the charge of one half cycle at high frequencies.

Characteristic frequency. At the characteristic frequency, ω_{RC}, the current is divided into equal parts on both branches of the RC circuit. The phase angle θ is $45°$ and the imaginary part reaches a minimum. This is visible in the plot of the negative imaginary part of the impedance over the logarithmic frequency as a clear peak in figure 2.11(b). The real part of the impedance shows the steepest slope in the corresponding diagram at this frequency.

---

[9]The angular frequency ω should not be confused with the frequency f, which is the inverse of the time period T. Both frequencies are connected via $\omega = 2\pi f$.

[10]The (incremental) capacity that can be observed in LIBs plays a special role in this context, which will not be discussed here, but is explained in [27, 28].

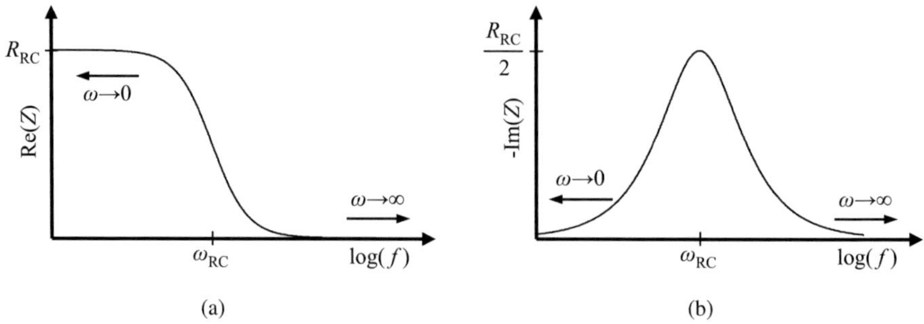

Figure 2.11.: (a) Real part and (b) negative imaginary part of the impedance of a RC circuit plotted over the logarithmic frequency.

Low frequencies. For lower frequencies, the charge transported during one half cycle increases. The charge that can be accepted by the capacitor remains the same, assuming voltage excitation. The charge stored on the capacitor plays only a minor role in comparison to the charge that is moved in one half cycle and the impedance approaches R_{RC}.

Very low frequencies. When the frequency approaches zero, the charge transported approaches infinity and the capacity of the capacitor is negligible compared to the total charge and the ohmic behavior is predominant. The imaginary part is zero again and the real part has a value of R_{RC}.

Characteristic curve. The shape of the whole impedance curve of an RC circuit is a perfect half circle. For this reason, electrochemical impedance data should always be plotted with equally scaled axes in a Nyquist plot. Only in this way can the data be visually determined to ideal (RC circuit) or non-ideal behavior (one of the impedance elements introduced in section 2.3.3.5).

Analogy to electrochemical systems. The most prominent electrochemical analogy to an RC circuit is a charge transfer process associated with a double layer capacity that corresponds to the capacitor in the RC circuit. The charge transfer itself can be regarded as an ohmic resistor, because it is associated with a certain loss, that is proportional to the species transferred.

2.3.3.4. Ohmic Losses in Electrochemical Systems

Ohmic losses occur whenever current flows through an electrochemical system because every component exhibits a finite conductivity, with the exception of superconductors. In this thesis the ohmic losses are summarized as $\eta_\Omega = J \cdot R_\Omega$.

The majority of the ohmic resistance is typically caused by the electrolyte. While good electronic conductors are relatively easy to find, good ionic or proton conductors that exhibit a negligible electronic conductivity are rare and the achievable corresponding conductivity is rather small.

The electronic conductivity of the electrodes is generally quite high compared to the electrolyte. This is achieved by using good electronic conductors for the electrode itself (like

Lanthanum Strontium Cobalt Ferrite (LSCF), a SOFC cathode material), utilizing an electronic conducting phase in the electrode (like the Ni in a SOFC anode), or by adding an electronic conducting material to the electrode (like carbon black in LIB cathodes).

A further contribution to the ohmic resistance originates from contact resistances between cell components or from the formation of undesired interlayers as the following examples demonstrate:

- It was shown in [27], that a significant part of the ohmic resistance in a LIB test cell originates from the large contact resistance between electrode and current collector.

- The formation of an insulating layer of Lanthanum Zirconia (LZO) during cell operation may increase the ohmic resistance of a SOFC with a Lanthanum Strontium Manganite (LSM) cathode as shown in [29, 30].

It should be mentioned here, that the impedance response of the Yttria Stabilized Zirconia (YSZ) electrolyte in SOFCs is assumed to be purely ohmic and this fact is widely accepted. However, it only appears as ohmic resistance in impedance studies, because the relaxation frequency at operating temperatures above $T = 600\,°C$ is $f > 10\,\mathrm{MHz}$ and is therefore not measurable with standard measurement equipment (see also section 3.5.1.6). This is because, that also the hopping mechanism between different lattice sites responsible for ionic conductivity occurs at a very high rate [31]. If the operating temperature is decreased significantly, the impedance response of the electrolyte exhibits a polarization behavior allowing for bulk and grain boundary contributions to be analyzed separately [31, 32]. Polarization processes are subject to the following section 2.3.3.5.

2.3.3.5. Polarization Losses in Electrochemical Systems

Losses occurring in an electrochemical system that are not purely ohmic are the so-called polarization losses. This behavior can be observed in many technical processes and is best explained by an RC circuit, its electronic equivalent. In fact, all electrochemical processes can be broken down to a combination of RC circuits as previously stated in section 2.3.3.3.

However, a number of more complex circuit elements has been established as a standard to describe the behavior of electrochemical systems. A selection of special equivalent circuit elements relevant for this thesis will be introduced in the following subsections. For further reading, the help function of *Zview* is highly recommended [33].

RQ Circuit

The RQ circuit can be interpreted as a generalization of the RC circuit when using the following representation:

$$Z_{\mathrm{RQ}}(\omega) = \frac{R_{\mathrm{RQ}}}{1 + (j\omega\tau_{\mathrm{RQ}})^{\alpha}}, \quad \alpha = 0\ldots 1. \tag{2.22}$$

However, more common is the following representation employing the Constant Phase Element (CPE) or Q element [21]:

$$Z_{\mathrm{Q}} = \frac{1}{(j\omega)^{\alpha}Y_0}, \quad \alpha = 0\ldots 1. \tag{2.23}$$

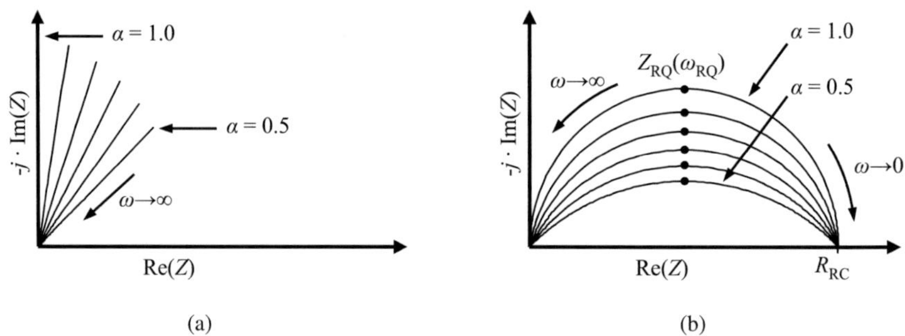

Figure 2.12.: Nyquist plots of (a) a Q element and (b) an RQ circuit with varying values for α, respectively.

The Q element has no direct electrical equivalent. A physical interpretation is proposed in [34]. For $\alpha = 1$, the Q element equals a capacitor and the RQ circuit equals an RC circuit, as indicated by the Nyquist plots of the Q element and the RQ circuit in figure 2.12. The corresponding description is

$$Z_{\mathrm{RQ}}(\omega) = \frac{R_{\mathrm{RQ}}}{1 + (j\omega)^\alpha R_{\mathrm{RQ}} Y_{\mathrm{RQ}}}, \quad \alpha = 0 \ldots 1. \tag{2.24}$$

The RQ circuit was first introduced to approximate the bioimpedance of a human body [35]. The measurable impedance in a human body is determined by the impedance of fat cells and muscle cells and cannot be approximated by a perfect RC circuit. Instead, it is more accurately modeled by a composition of parallel and serial connections of them, leading to a flattened semicircle in the Nyquist plot [18, 36]. In the same way that a body is composed of cells with varied responses, electrodes in an electrochemical system are composed of different impedance responses for different grains or reaction sites, that vary slightly in parameters and cannot assumed as ideal.

For an RQ element, a single relaxation time, τ, cannot be determined. This is explained in more detail in section 4.4. Due to the fact that the exponent in equations 2.22 and 2.24 is not an integer value, simulation in the time domain is not possible directly [37]. An approach to obtain an analytical expression in the time domain is proposed in [37]. Additionally a time domain simulation can be conducted by the software tool *plecs* in *Matlab*. As an alternative an RQ circuit was approximated by a finite number of serial RC circuits by [38].

A problem concerning the RQ element is the physical interpretation. In the applied Lorentz distribution, the parameter α does not have a physical meaning as discussed in [35]. Some authors argue, that the Lorentz distribution should be substituted by a Gaussian distribution, which would yield a direct relation of the parameter α to the standard deviation, σ^2, of the relaxation time, τ [35]. Nevertheless, the frequency domain expression for a Gauss element is somewhat more complex and more difficult to fit to experimental data.

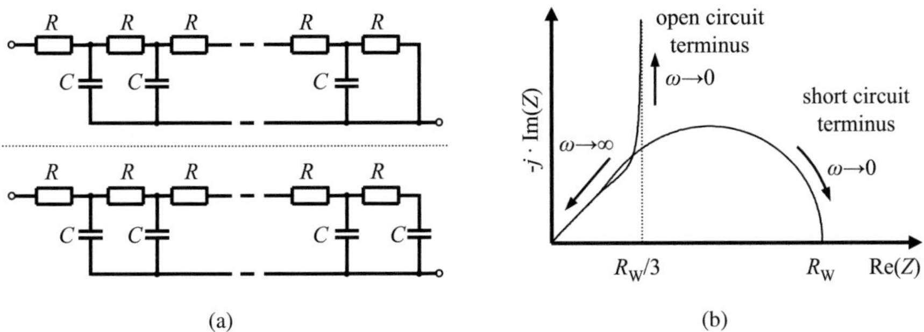

Figure 2.13.: (a) Equivalent circuit composed of basic impedance elements for the Generalised Finite Length Warburg element (Short Circuit Terminus and Open Circuit Terminus) and (b) Nyquist plots of both Warburg elements.

Warburg Element

An important impedance element is the Warburg element. The correct title of the version used in this thesis is: generalized finite length Warburg element (short circuit terminus). Its impedance expression is

$$Z_{\mathrm{W}}(\omega) = R_{\mathrm{W}} \cdot \frac{\tanh{(j\omega\tau_{\mathrm{W}})^{\alpha_{\mathrm{W}}}}}{(j\omega\tau_{\mathrm{W}})^{\alpha_{\mathrm{W}}}}. \tag{2.25}$$

This element is used to model gas diffusion in the anode substrate of an SOFC in [39]. It is discussed in more detail in [18]. The ECM with basic impedance elements and its associated Nyquist plot are shown in figure 2.13.

The physical interpretation of this behavior is that this circuit represents the diffusion length. In order to pass a length element, a certain amount of work is necessary (modeled by the resistors). In addition to this, some overconcentration can be stored in each length element (modeled by the capacitor). The short circuit on the upper branch signifies that this whole branch has the same electrical potential. This is sensible since SOFC anodes exhibit good electronic conductivity. The short circuit on the right end represents an endless supply of diffusing species that is related to the constantly supplied gas in the gas channel.

Note that there is a version of the generalized finite length Warburg element, called the open circuit terminus, that is sometimes used in LIB analysis to represent the limited species in the electrodes indicated by a capacitor on the right end. Its ECM using basic impedance elements is also shown in figure 2.13(a).

The impedance curves for both versions of the Warburg element are presented in figure 2.13(b) and show the characteristic slope of 45° for high frequencies and switch to a semicircle for lower frequencies. The open circuit terminus behaves like the serial connection of a short circuit terminus Warburg element and a capacitor and switches into a straight capacitive branch for low frequencies.

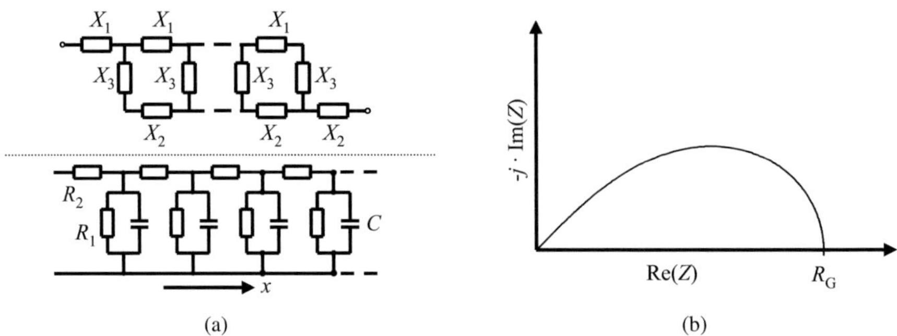

Figure 2.14.: (a, top) General TLM with an electronic conducting path (upper branch, X_1), an ionic conducting path (lower branch, X_2) and the corresponding charge transfer impedance X_3 and (a, bottom) equivalent circuit composed of basic impedance elements for the Gerischer element and (b) typical Nyquist plot of a Gerischer element.

Transmission Line Model

Like the Warburg element, the general Transmission Line Model (TLM) represents the spatial distribution of the electrochemical processes occurring in an electrode. Figure 2.14(a) shows a general example of such a TLM with an electronic conducting path (upper branch, X_1), an ionic conducting path (lower branch, X_2) and the corresponding charge transfer impedance, X_3. This element is also referred to as Bisquert element. A detailed description can be found in the literature [40, 41].

Gerischer Element

The Gerischer element is the result of a infinitely long TLM with negligible electronic losses in the upper branch, as shown in figure 2.14(a). In this case, the analytical expression converges to

$$Z_{\mathrm{G}}(\omega) = \frac{R_{\mathrm{G}}}{\sqrt{1 + j\omega\tau_{\mathrm{G}}}}. \tag{2.26}$$

This is explained in further detail in [42]. The Gerischer element is typically applied for SOFC cathodes [10]. Also the widely used Adler model for cathodes is based on a Gerischer element [43]. For further information about cathode models see also [10]. An example for the Nyquist plot of a Gerischer element is shown in figure 2.14(b).

2.3.3.6. Remarks on the Impedance of Electrochemical Systems

Three remarks should be added to this section, in order to complete the overview of the impedance of electrochemical system of fuel cells and batteries.

Ohmic and Polarization Resistance

In the literature the terms R_0 and R_{pol} are often used. They are defined together with Z_{DC}, as demonstrated in figure 2.15(a). The ohmic resistance R_0 is the high frequency

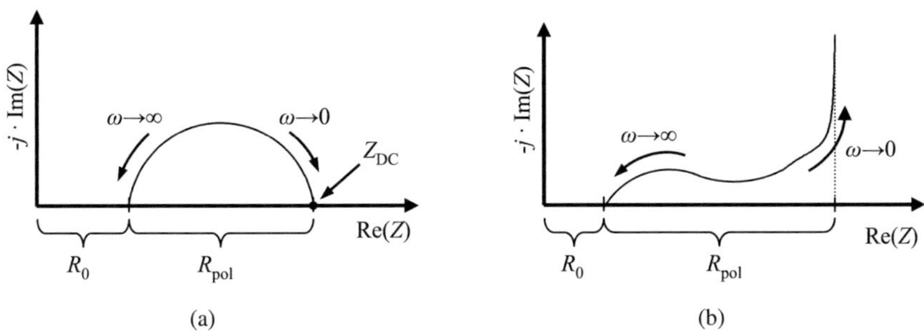

Figure 2.15.: (a) Scheme of a Nyquist plot indicating R_0, R_{pol} and Z_{DC} and (b) a typical Nyquist plot of the impedance of a Li-ion battery.

intersection of the Nyquist curve of the impedance with the real axis and Z_{DC} is its low frequency intersection with the real axis (not existent for LIBs, see below). The polarization resistance is defined as $R_{\text{pol}} = Z_{\text{DC}} - R_0$.

Relation between C/V Characteristics and Impedance Curves for SOFCs

C/V curves represent the static behavior of an electrochemical system at various potentials, whereas an impedance spectrum represents the dynamics of it at a fixed potential. The correlation between these two characteristics is the DC resistance Z_{DC}. For SOFCs Z_{DC} is given as the impedance for $\omega \to 0$ and is purely ohmic. It depends on the current density and is also represented by the slope of the C/V curve at a given current density, as indicated in figure 2.5. If both characteristics are analyzed for an electrochemical system, the two methods can be checked for consistency by comparing the DC resistance Z_{DC} obtained by the two methods. For LIBs this correlation is not so simple, as will be discussed below.

Capacitive Branch of LIBs

For low frequencies, batteries show capacitive behavior. This means, as $\omega \to 0$, the imaginary part is a complex number with large negative imaginary part (cf. equation (2.19) and figure 2.15(b)). This behavior can be explained by the OCV curve of the battery, which increases, as the cell is charged, and decreases, as the cell is discharged. The linear approximation of this behavior in the operating point is a capacitor. For this reason, a large number of battery models introduce a serial capacitor to account for this capacitive behavior of LIBs.

The definition of the polarization resistance is not straightforward for LIBs. However, this issue is omitted here, as the polarization resistance of LIBs will not be needed in this thesis.

2.3.4. Performance and Efficiency

An important goal of current research activities is to get impartial numbers for the performance and efficiency of an electrochemical system. This is one of the major purposes of the models presented in chapter 5. Possible criteria to rate performance and efficiency are controversially applied throughout the literature with little consistency.

Basically, the performance of an electrochemical system is determined by the physical quantities energy density and power (loss) density. Efficiency is the ratio between energy delivered by the system and energy supplied to the system. The analysis differs fundamentally for fuel cells and batteries, and is introduced separately below.

2.3.4.1. Fuel Cells

A fuel cell produces electrical and thermal energy by the conversion of fuel. The major issues completing a performance and efficiency analysis are the quantification of the individual loss processes. Based on the power (loss) density, adequate values can be derived at cell level and system level.

Power (Loss) Density as Performance Criterion

If the individual loss processes during operation of an electrochemical system are to be identified quantitatively, an adequate measure has to be chosen first. As pointed out in section 2.3.2, there are different definitions for the resistance and impedance for electrochemical systems in the literature. The same is true for performance and efficiency. A common way to specify the performance of a fuel cell is to report the power density at a certain operating point, normally defined by temperature, anode and cathode gas concentrations and operating voltage. But at high current densities, there is also an influence of the gas flow rate, if the gas conversion over the cell area is not negligible, as will be demonstrated in chapter 5. Additionally, the size of the cell must be taken into account for performance analysis.

In this thesis the power density P and the power loss density P_{loss} in the operation point are proposed as good criteria for evaluating the performance of a fuel cell. The cell voltage $U_{\text{op}}(J)$ is relatively easy to measure. The output power density is:

$$P_{\text{tot}} = U_{\text{op}}(J) \cdot J. \tag{2.27}$$

Of further interest are the losses, defined as total power loss density $P_{\text{loss,tot}}$:

$$P_{\text{loss,tot}} = \eta_{\text{tot}}(J) \cdot J = (U_{\text{OCV}} - U_{\text{op}}(J)) \cdot J. \tag{2.28}$$

The total overvoltage can be separated into contributions of the individual cell components. The overpotential η_k, for a specific process k, is needed for this task. If it can be determined as a function of the current density, the corresponding power loss density follows as

$$P_{\text{loss},k} = \eta_k(J) \cdot J. \tag{2.29}$$

The derivation of each η_k is not trivial, as a nonlinear system has to be considered (cf. section 2.3.2). Possible ways to identify $\eta_k(J)$ will be described in sections 5.3.2 and 5.3.5 for small-scale and large-scale systems, respectively.

Cell Efficiency

The term

$$\eta_{\text{el}} = \frac{P_{\text{loss}}}{P + P_{\text{loss}}} = \frac{U_{\text{op}}}{U_{\text{OCV}}} = -\frac{zFU_{\text{op}}}{\Delta G} \tag{2.30}$$

yields the electrochemical efficiency of a cell (cf. [14] and equation (2.11)). This states that the efficiency of a fuel cell is defined as the ratio of the usable electrical energy produced by the fuel cell and the chemical energy inserted, because the fuel cell is regarded as an converter of usable chemical energy ΔG to electrical energy (see also section 2.4.2). Heat is also produced during operation by a fuel cell, but heat generation will only be taken into account at the system level and not as part of the cell efficiency.

System Efficiency

Heat produced by a fuel cell system may be utilized in applications such as Combined Heat and Power (CHP) systems, so it is sensible to specify also the ratio between thermal energy produced by the fuel cell and the chemical energy inserted. This is referred to as thermal efficiency η_{th} [14]. Electrical and thermal efficiency are often summed up to the overall efficiency, $\eta_{\text{el}} + \eta_{\text{th}} = \eta_{\text{tot}}$. As the losses created by the overpotentials η_k, that lower the cell efficiency η_{el}, are dissipated as heat and consequently are added to the thermal efficiency η_{th}, η_{tot} can come close to 1.

A very high electrical efficiency of 68% for SOFC systems was presented in [44]. It has to be noted that such high efficiencies are only possible for low overpotentials. These can only be realized with relatively small current densities which result in low system power and energy densities.

Energy Density

On the cell level, the energy density of a fuel cell is not a relevant number because it is a compound of thin layers to which the fuel is supplied. On the system level, energy density becomes a more sensible variable to compare performance, but due to the variety of fuel cell types and geometries, such as tubular and planar for SOFCs, no general numbers can be provided here.

2.3.4.2. Batteries

Batteries are rated according to capacity and recommended C rate. These characteristics can be broken down to energy density and performance plus efficiency, respectively.

Energy Density

A high energy density is the most important performance related specification for a favorable electrode material for LIBs. The energy density can be expressed per unit volume or weight, depending on what is more relevant for a given application. Performance itself can be controlled to a certain extent by the micro- and macrostructure of the cell.

High Energy and High Power Cells

A battery cell is composed of the cell components introduced in section 2.2.1. In this paragraph it is briefly demonstrated, how the composition of a cell can be trimmed for high energy density and high power density. The measures that can be taken are explained for the three relevant cell components:

- **Separator with electrolyte.** The thickness of the electrolyte does not contribute to storage and is reciprocal to the ohmic resistance of the cell. Therefore the electrolyte is as thin as possible to achieve both higher power and energy densities.

- **Current collectors.** The thicker the current collectors, the lower the resistance caused by them. In high power cells, thick current collectors are used in order to increase the maximum current and decrease heat dissipation. However, as the thickness of the current collector increases, the thickness of the electrodes must decrease for equally sized cells. As a consequence of having thick current collectors, high power cells have less system energy density than other cells.

- **Electrodes.** The thicker the electrodes the fewer number of layers is needed to achieve a given capacity. For this reason, high energy cells commonly have thicker electrodes than other cell types and the volume ratio between electrodes and current collectors in the overall volume is therefore higher. Due to few and thin current collectors, the maximum current is limited for this type of cells.

Efficiency

For batteries the efficiency is given by the ratio of energy that is needed to charge the battery and the energy that can be used in the outer electrical circuit during discharge:

$$\eta_\mathrm{C} = \frac{E_\mathrm{discharge}}{E_\mathrm{charge}}. \tag{2.31}$$

The efficiency typically depends on the C rate at which the charge and discharge cycle is performed, but even at very low C rates, the voltage is higher during charging than during discharging at the same SOC (cf. figure 2.7(a)). This is mainly due to overpotentials and the hysteresis in LIBs[11]. At high C rates the efficiency drops, because at higher currents, overpotentials rise and are thereby lowering the efficiency.

2.4. Thermal Properties of Electrochemical Systems

Temperature is a very important operating parameter for electrochemical systems. There are several facts, that underline this importance:

- The OCV is strongly dependent on the temperature.

- Almost all loss processes depend strongly on the temperature.

---

[11]The hysteresis is the observable difference in the OCV at the same SOC in steady-state, if the cell has been charged or discharged to the present SOC. It is an important phenomenon for LIBs, as can be seen from recent publications like [45] and [46], for example, but will not discussed in further detail in this thesis.

- Losses produced by the loss processes themselves are dissipated as heat.

- Heat is generated or consumed by the electrochemical reactions during normal operation.

- The systems have a defined temperature window, in which safe, stable and efficient operation is possible.

The information relevant to this thesis is prepared in two sections, the first one dealing with possibilities and limitations concerning the operating temperature (section 2.4.1) and the second providing the equations needed to calculate heat dissipation for a given electrochemical system (section 2.4.2).

2.4.1. Operating Temperature

In general, a higher operating temperature will lead to more efficient operation for the here discussed electrochemical systems. This is primarily due to the strong temperature dependency of the electrolyte conductivity. In [31] the conductivity of the standard electrolyte material YSZ is studied in detail. In [47] some conductivities of electrolytes for LIBs are listed. Many electrode reactions are temperature activated, but it is not so easy to give a general statement here.

It should be noted here, that even though the Gibbs free energy for the electrochemical reaction in SOFCs decreases with increasing operating temperature resulting in a lower OCV, the performance of SOFCs increases with increasing temperature at technically relevant operating points.

While the operating temperature affects many electrochemical processes and is an important testing parameter, most systems have a tightly defined temperature window for normal operation:

- **Stability of materials.** In general, high operating temperatures involve higher degradation rates. Additionally, mechanical stress due to mismatched thermal expansion coefficients of applied materials is an issue, which can lead to system failure.

- **Carbonization (SOFC).** SOFCs should not be operated at too low temperatures, if there are carbonaceous gases contained in the fuel gas, like is the case for diesel reformat or biogas. Depending on the catalytic activity and the humidity the anode experiences, the risk of carbonization rises with a decrease in operating temperature below 650°C [48].

- **Safety.** Thermal runaway of rechargeable batteries is serious safety problem. High temperatures and overcharge can trigger runaway exothermal chemical reactions, that may lead to combustion or even explosion [49]. This is why temperature monitoring provided by a Battery Management System (BMS) is essential.

2.4.2. Heat Dissipation

Heat dissipated during operation is an important physical quantity. Heat dissipation is a complex topic, because heat produced during operation can be beneficial, as it is in SOFCs to maintain operating temperature, or detrimental in the case of LIBs because of the risk of thermal runaway.

Particularly for LIBs and SOECs it is important to introduce the three different heat sources or sinks in an electrochemical system. All of them are related to the current density J, which means, that no heat is emitted or consumed under equilibrium conditions.

- **The electrochemical reaction.** The overall reaction enthalpy ΔH is divided into two parts: $\Delta H = \Delta G + T \Delta S$. In the case of an exothermal reaction, the amount of $T \Delta S$ is directly given as heat to the surrounding, ΔS being the reaction entropy. In case of an endothermal reaction, $T \Delta S$ represents the heat sink. The free molar enthalpy ΔG is turned into electrical energy with the efficiency η_{el}, which can be utilized in the external electric circuit.

- **Voltage drop at the electrolyte.** Due to the ohmic resistance of the electrolyte, there is a voltage drop at the electrolyte and joulean heat is emitted. Its value is calculated by $P_{\text{J}} = J^2 R_\Omega$.

- **Voltage drop at the electrodes.** The electrodes act as heat sources, when an electric load is applied. Their contribution being $P_\eta = J \sum \eta_{\text{tot}}(J)$.

The overall heat emission is then calculated by ([16])

$$
\begin{aligned}
P_{\text{tot}} &= J \frac{T \Delta S}{zF} + J \sum \eta_{\text{tot}}(J) + J^2 R_\Omega & (2.32) \\
&= J\left(U_{\text{H}} - U_{\text{OCV}}\right) + J \sum \eta_{\text{tot}}(J) + J^2 R_\Omega. & (2.33)
\end{aligned}
$$

The first summand of equations (2.32) and (2.33) shows a linear dependency on the current density J. The second summand exhibits nonlinear behavior due to the nonlinear dependency of the overpotentials on the current density. However, this behavior can often be approximated by a linear function which allows the second summand to be approximated as a quadratic function. The third summand equals a quadratic function by definition. In conclusion, the emitted heat shows a quadratic dependency on the current density because of losses at the electrodes and electrolyte.

The thermal cell voltage $U_{\text{H}} = -\Delta H / zF$ is explained in detail in [16]. The values necessary for the evaluation of equations (2.32) and (2.33) can be found in [14, 16] for several fuel cell types and in [6] for a large number of battery types.

SOFC

For SOFCs in H_2/H_2O in operation all terms of equations (2.32) and (2.33) are positive for all current densities.

For SOFC operation utilizing reformat as fuel, a number of endothermal and exothermal reactions are likely to take place in parallel depending on the applied reformat. For this reason ΔS is not easily determined. In [48] SOFCs in reformat operation are studied in detail.

SOEC

If an SOFC is operated in electrolysis mode, ΔS is negative and the cell acts as a heat sink for small current densities. There is a so-called thermoneutral point, when the heat emitted through losses is equal to the heat consumed by the electrochemical reaction. This will be further discussed in section 3.5.3.1.

LIB

Determining the heat emission of LIBs is more challenging because of the presence of different and evolving materials, phases and reactions in both charge and discharge mode. The discussion of heat dissipation is of greater importance because of the limited temperature window for safe and efficient operation. In [50, 51] methods have been developed to analyze the heat emission.

2.5. Chemical Properties of Electrochemical Systems

Although the focus of this thesis is on measurement and modeling of electrochemical systems, chemical properties of the system under test need to be introduced. This section is divided into two parts: in section 2.5.1 applied materials are introduced and in section 2.5.2 some examples for the stability limits of components are given. As these limits are correlated to the electrical potential applied, they may be detected or even triggered via the application of current or voltage and therefore makes it an interesting topic for the measurement techniques presented in this thesis.

2.5.1. Materials

The standard material configuration for the SOFCs considered in this thesis is stated and explained in this section, accompanied by a short overview of recent developments in materials research. A brief overview of LIB materials is also provided in section 2.5.1.3.

2.5.1.1. Materials for SOFCs

The range of material systems used for SOFCs cannot be enumerated. In order to provide the most general of examples, the cells measured in this study are standard materials at the time of writing this thesis. These materials are briefly introduced together with recent developments in the following subsections. The actual cell configuration for the measurements in this thesis is provided after the overview of the materials for anode, electrolyte and cathode. It should be noted that the measured cells were either half cells on which the cathode was deposited at IWE or delivered as full cells by Forschungszentrum Jülich. A list of the tested samples can be found in the appendix (section A).

Anode

More than 40 years ago, Spacil's patent introduced fine Ni particles homogeneously distributed in a YSZ matrix [52]. Since then, numerous quantitative models related anode microstructure to performance, and gave advice for an optimum connectivity of gaseous, ionic (YSZ) and electronic (Ni) phases [53, 54]. Other experimental studies documented outstanding initial performance of custom tailored Ni/YSZ composite anodes inevitably followed by degradation in long-term operation and/or high fuel utilization (high p_{H_2O}) testing conditions [55, 56, 57, 58]. Qualitative correlations between microstructure and power density during operation identified coarsening and agglomeration of Ni particles [53] as the main source of degradation, thus reducing the Three Phase Boundary (TPB) length. This source of degradation cannot be avoided and even fully developed manufacturing techniques only slow down, but do not stop anode degradation at constant load. This effect is more pronounced over repeated re-oxidation cycles [59]. Research has shifted to metal-oxide based anodes, such as doped $SrTiO_3$, which are less susceptible to microstructural changes. However, adequate power densities are only reported when a catalytically active material, such as Ni [60] and Pd [61], is added.

Electrolyte

The standard material used for the electrolyte in SOFCs is YSZ. It shows a high ionic conductivity due to the creation of mobile oxygen vacancies when ZrO_2 is doped with Y_2O_3 [62]. The most commonly used stoichiometry is $Zr_{0.83}Y_{0.17}O_{2-\delta}$ (8YSZ), corresponding to 8 mol% Y_2O_3-doped ZrO_2. The crystal structure of zirconia is stabilized to a cubic fluorite crystal structure, but it should be mentioned, that recently, there have been some discussions about this stability under reducing conditions due to indiffused Ni during co-sintering of anode and electrolyte [62]. The electronic conductivity of this material is negligible.

Although the term YSZ is not precise as different dopant concentrations of YSZ exist, the term YSZ will represent this composition in this thesis. For a more detailed introduction to YSZ and alternative electrolyte materials, the reader is referred to [31].

Cathode

The cathode material used for the cells in this thesis is LSCF. The chosen stoichiometry is $La_{0.58}Sr_{0.4}Co_{0.2}Fe_{0.8}O_{3-\delta}$. It is a Mixed Ionic Electronic Conducting (MIEC) material, which means that TPB points cannot only be found at the interface cathode/electrolyte but across the surface of the cathode [63]. Details about the optimization of the stoichiometry can also be found in [63]. The initial performance of this cathode is very good, however, there are issues about the chemical compatibility with YSZ. Therefore, a Gadolinium Doped Ceria (GDC) interlayer has to be applied between LSCF and YSZ [64]. This interlayer retards the formation of an insulating interlayer consisting of $SrZrO_3$ between cathode and YSZ [65].

Another issue is a pronounced degradation, which has been analyzed during the first 1000 hours of operation for various operating temperatures in [66].

Other cathode materials that should be mentioned are the widely used LSM/YSZ composite electrode [63, 67] and Lanthanum Strontium Cobaltate (LSC), which exhibits very promising performance values as nano-cathode for operating temperatures of 600°C or below [10].

Assembly

SOFC measurements in this work were performed on planar Anode Supported Cells (ASCs), based on an anode substrate (Ni/YSZ, thickness $0.5\ldots1.5$ mm). The Anode Functional Layer (AFL) (Ni/YSZ, $d = 13\,\mu$m) and the electrolyte (YSZ, $d = 10\,\mu$m) were deposited subsequently by vacuum slip casting. On top of the electrolyte, a Gd_2O_3-doped CeO_2 (GDC) interlayer was screen-printed and sintered at $T = 1300\,°C$ for $t = 3$ h, resulting in a thickness of $d = 7\,\mu$m [64]. On top of this interlayer, a LSCF cathode was applied via screen-printing and sintered at $T = 1080\,°C$ for $t = 3$ h, resulting in a thickness of $d = 45\,\mu$m with different sizes (see section 3.2.1 and figure 3.1). On some of the cells, two reference electrodes were screen-printed both before the working cathode and behind it (in direction of the gas flow), in order to monitor the open circuit voltage. Anode substrate had a cross-sectional area of $A_S = 50 \times 50$ mm^2 and were entirely covered by the thin-film electrolyte. Details on cell fabrication technology are described in detail in [68].

Electrical Contact

The experiments in this thesis were all conducted with ideal electrical contacting (ideal current collectors). A Ni mesh is used on the anode side where a Ni phase is present in the cell and gives good electrical conductivity. Au is used on the cathode side because the noble metal is stable in oxidizing conditions and is chemically compatible with the cathode material, LSCF.

Stack Design

The corresponding stack design, for which the test cells used in this thesis were developed, is explained in detail in [69]. New developments are presented in [70]. Changes in performance data due to non-ideal electrical contacting in the stack is measured and modeled in great detail in [67].

2.5.1.2. Materials for SOEC

A SOEC is just a reversely operated SOFC. In order for systems to be eligible for both operation modes, the same materials are used for SOECs as for SOFCs. The SOEC experiments reported in this thesis were conducted on the same cells as described for SOFCs.

2.5.1.3. Materials for LIBs

The material systems used for cell components in LIBs is even more numerous than for SOFCs. This applies to basic materials, material compositions, and fabrication. Further, the materials used are coated, blended, and mechanically processed in different ways, so that it is difficult to give a quick overview here.

Materials are chosen due to their electronegativity or potential versus a reference electrode, which should be high for cathode materials and low for anode materials, because the difference in standard potentials is directly linked to the cell voltage. The standard potentials versus Li can be found in [71], for example. A good overview of battery materials is also given in [72]. Recent developments, characteristics and prices are compared in

[73]. A detailed impedance study on LIBs with a graphite anode and a LiFePO$_4$ cathode can be found in [27]. In [74] a Focused Ion Beam (FIB)/Scanning Electron Microscopy (SEM) reconstruction of a LiFePO$_4$ cathode is presented, enabling the identification of the microstructure parameters for LIB cathodes.

The LIB cells measured in this thesis were commercial cells with a capacity of $Q = 120\,\text{mAh}$. They consist of a LiCoO$_2$ cathode and a graphite anode. The protection circuit module was removed before testing. As no further details about the applied materials are given by the cell provider (*Reva*) and the components were not further investigated, the reader is referred to the references above for an overview and characteristics of common materials for LIBs.

2.5.2. Some Remarks on Chemical Stability

As pointed out in the previous section, materials play a large role in electrochemical systems and with great effort they have been optimized for the corresponding purpose. However, due to harsh operating conditions and large overpotentials, these materials may be oxidized, reduced, decomposed, dissolved or may precipitate during operation.

In this section four such scenarios will be explained in detail. In general, every material has a chemical stability window. At a neutral potential, this can be expressed by an oxygen partial pressure above or below which a material is oxidized or reduced, respectively. The chemical stability limit strongly depends on temperature. If an overpotential is present in an electrochemical system, the oxygen partial pressure is altered because of the correlation with the electrochemical potential $\bar{\mu}$, introduced in equation (2.5). An accurate determination of the oxygen partial pressure present in the different active layers in an electrochemical system is quite complicated. The chemical stability of a material is commonly tested in simple model setups like symmetrical electrodes or accurately specified small geometries via Cyclic Voltammetry (CV) (see section 3.4). The following examples constitute the theoretical background behind some practical examples.

2.5.2.1. Stability of the Nickel Anode in SOFC

The standard anode for SOFCs consists of Ni and YSZ (see section 2.5.1.1). Ni is stable at high temperatures under reducing conditions. The fuel gas on the anode side has a reducing character[12]: a high cell voltage is achieved by a low oxygen partial pressure on the anode side (reducing) and a high oxygen partial pressure on the cathode side (oxidizing)[13] (see equation (2.12)).

There is a clear stability limit for the oxygen partial pressure depending on temperature, above which metallic nickel reacts with oxygen to nickel oxide [75, 76, 77].

This reaction is reversible, however, the phase change from Ni to NiO involves a volumetric expansion of 69 % [78] and the growth and subsequent shrinkage of the nickel particles in the cermet anode cause irreversible changes to the microstructure of the anode [77]. After a few so-called 'redox-cycles' the ASR can even be decreased by coarsening

---

[12]In case of humidified H$_2$, the oxygen partial pressure ranges between $p_{O_2} = 10^{-24} \ldots 10^{-18}$ atm.

[13]Air is used as cathode gas. The corresponding oxygen partial pressure is $p_{O_2} = 0.21$ atm.

of the anode due to micro-cracks. Nevertheless, repeated redox-cycles lead to Ni agglomeration and decrease the mechanical stability of the anode structure, which can result in cell cracks and cell failure in the worst case.

There are several reasons for the oxygen partial pressure to rise over the critical limit during SOFC operation:

- accidental breakdown of fuel supply;

- warm start of a fuel cell system with reformer when the fuel cell stack is already at operating temperature but the reformer is not ready yet to provide the necessary fuel gas;

- a bad operation point of the fuel cell stack with a high p_{O_2} (high p_{H_2O}, see equation (2.12) and (2.13)) at the end of the gas channel, induced by a too large current density applied.

One special case for the damage of the SOFC anode by redox cycles was studied in a research project ('Herausforderung Brennstoffzelle' [79]), where the impact of the inverter feedback on the stability of the cell was investigated. Inverters are used to connect electrochemical systems (DC voltage sources) to the Alternating Current (AC) power network. Their load signal is characterized by an offset current plus a 100 Hz ripple (double AC frequency of the power network). It was found, that a large 100 Hz ripple can cause superficial redox-cycles in the case of a low operation voltage ($U_{op} \leq 0.7$ V) at the gas outlet of a SOFC stack, where a high p_{H_2O} is present in the fuel gas in order to achieve high power densities.

2.5.2.2. Stability of the YSZ Electrolyte in SOFC

The standard material for SOFC electrolytes is YSZ (see section 2.5.1.1). It is stable at all relevant operating conditions for fuel cell operation. If the cell is operated under dry conditions (low p_{H_2O} at the fuel electrode) in SOEC mode at high current densities, the potential on the interface electrolyte/anode can reach a critical value, so that the zirconia is reduced to zirconium [80].

However, it was found, that the re-oxidized boundary layer has better electrochemical properties than the structure had before, due to an altered microstructure as shown in [81, 82, 83, 84]. The microstructural change after re-oxidation can be observed in figure 2.16. The new structure was investigated by SEM and Transmission Electron Microscopy (TEM) and showed a 200 nm thick nano-structured interlayer consisting mainly of re-oxidized YSZ between anode and electrolyte, that is likely to have increased the electrochemically active area of the anode. Because of the beneficial impact on cell performance, the technique of adjusting a short time reverse current density has been denominated Reverse Current Treatment (RCT).

The potential reduction of YSZ has been investigated by many researchers as reviewed in [85]. A complete reduction of the electrolyte in SOEC mode is prevented by an emerging electronic conductivity under these condition as pointed out in [86] and can be verified by the conductivity diagram in [87].

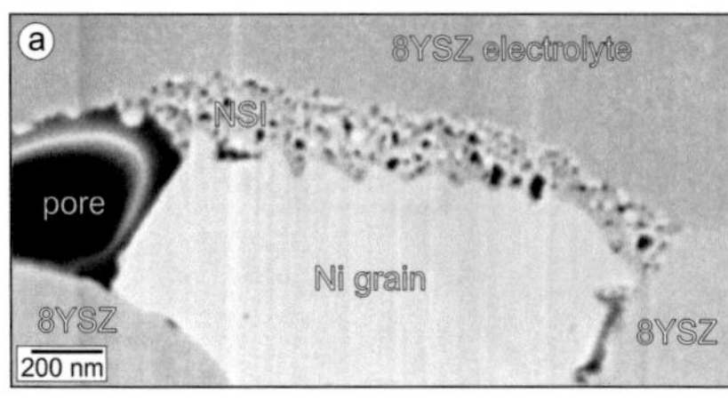

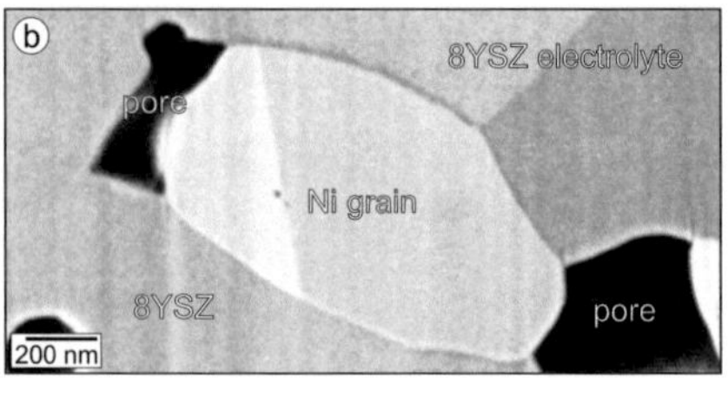

(a) (b)

Figure 2.16.: SEM images of the interface between AFL and YSZ electrolyte of sample S.1 after FIB polishing: YSZ electrolyte (top) and μm-sized Ni and YSZ grains within the AFL (bottom); (a) underneath the working electrode showing the nano-structured Ni/YSZ interlayer (NSI: center), which was formed by RCTs, (b) underneath the reference electrode and therefore not affected by the RCTs.

2.5.2.3. Li Plating in LIBs

Li plating is a serious problem for the success of LIBs. It occurs at low temperatures during charging of the battery at high C rates [88, 89].

During charging, Li ions are transported through the electrolyte to the anode, where they have to be inserted into the anode. At low temperatures the diffusion into the anode is very slow and causes large overpotentials. These can be so high, that a precipitation of metallic Li becomes energetically more attractive. Then Li dendrites are formed starting at the interface anode/electrolyte and growing towards the cathode.

This leads to a loss of active Li in the battery and can end up in an internal short circuit destroying the cell. For these reasons research currently focusses on possibilities for the early detection of Li plating.

2.5.2.4. Stability of the Electrolyte in LIBs

LIBs are attractive because of the high potential difference of possible anode and cathode materials. This results in a high cell voltage, which is directly linked to a higher energy density compared to systems with the same capacity but a lower cell voltage.

However, the high potential difference is responsible for demanding requirements for the electrolyte, as it has to show a high conductivity for Li ions and must be stable at all potentials present between anode and cathode.

This is a major drawback for the so-called '5 V battery' that aims to reach a cell voltage of $U_{\mathrm{op}} = 5\,\mathrm{V}$ [90]. As of the publication of this thesis, there is no standard electrolyte material available, that meets these requirements.

2.6. Modeling of Electrochemical Systems

The means to model an electrochemical system are chosen according to the model's purpose. As could be seen in the preceding sections, electrochemical systems are complex in their dependencies and interdependencies on various physical and chemical quantities. This introduces the first purpose of the intensive modeling activities for electrochemical systems during the last century:

- If a process can be characterized by a physically motivated model, materials and structures of electrochemical cells can be optimized, i.e. minimizing losses and maximizing performance and efficiency.

But this is not the only motivation for appropriate models of an electrochemical cell:

- For a complete description of an electrochemical system, all processes have to be identified. When all known basic relations of system processes can be modeled, the easier it is to identify previously unknown residual processes. This is was done in [91], where the accurate anode model developed in [39, 92] was applied to separate anode and cathode losses and to identify sources of long-term degradation in high performance SOFC cathodes.

- System models are commonly applied to Electronic Control Units (ECUs) including BMSs and diagnosis systems. The inclusion of these external components become necessary, as electrochemical cells move toward commercialization.

The main purposes of an ECUs are:

- to estimate the State of Health (SOH) of an electrochemical system,

- to act as a fault detection and diagnosis system, and

- to control the operating point of the system.

To be eligible for an ECU, it is a prerequisite, that the simulation in the time domain is possible with this model and that it can run with little computational effort on an ECU in an automobile or in a CHP system.

The choice for the right approach also depends on the size and the characteristics of the system. Therefore micro- and macro-models are distinguished in this thesis. But the focus in this section is the presentation of possible fundamental approaches in order to describe an electrochemical system: How different scales affect modeling methodology will be the subject of section 5.1.

Further, it is sensible to categorize the different approaches in static and dynamic models. For a static model all relevant quantities are calculated in one operation point. These models are needed to predict power output and efficiency in a system for example. Dynamic models try to reproduce the dynamic behavior of an electrochemical system and are used for identification, prediction and diagnosis of the whole system during operation but are also able to treat the system processes separately, as will be demonstrated in detail in section 5.3.

Realizations of the models vary in the same way the purposes for models vary. While physically motivated or physical models are preferable for the first two reasons above, they are hard to obtain and to parameterize in practice. Therefore alternatives have been established. The models can be classified into black box models, when no physical meaning is represented by the model parameters and the model is a pure behavior model. Due to the free choice of model equations, an accurate reproduction of the system behavior with respect to input and output variables is often achieved. White box model in contrast contain every known relationship between relevant variables. Between black box and white box models, literally every shade of grey is possible as a model description as pointed out in [93]. In the following section, some examples of white box (section 2.6.1), grey box (section 2.6.2) and black box (section 2.6.3) model approaches will be presented. A comprehensive overview of SOFC modeling activities is provided in [94].

2.6.1. White Box Models (Physical Modeling Approach)

The aim of these models is an exact physical description of the modeled system. Due to the inherent complexity of electrochemical systems, this goal is quite cumbersome. However, three approaches shall be introduced in this section that try to simulate electrochemical systems based on very fundamental physical and physicochemical relations.

2.6.1.1. Elementary Kinetics Modeling

These models try to reproduce the elementary kinetic steps in an electrochemical system [95], like the oxidation of CO on the TPB of Ni/YSZ patterned anodes [96] in order to apply these models to technically relevant structures. To accomplish this, homogenization (see section 2.6.2.2) is necessary because of computational complexity. This methodology results in a multiscale approach (see section 2.6.2.3).

2.6.1.2. Microstructure Modeling

The field of microstructure modeling is an emerging discipline in material science. Its aim is to establish geometric descriptions of real microstructures reproduced from test samples. On these microstructures, the fundamental thermodynamic and kinetic differential equations are applied to reproduce the static and dynamic behavior of the test sample during operation.

Owing to the availability of FIB/SEM devices, high resolution X-ray tomography devices and increasing computational power, the geometries of complex real electrode structures can be obtained and evaluated with multiphysics Finite Elements Method (FEM) software packages like *Comsol*. The progress during the last years led from small microstructures based on coarse artificial structures like [97] for SOFC electrodes to geometries that constitute Representative Volume Elements (RVEs) of real electrode structures [98, 99, 100]. The main drawback for microstructure modeling is model complexity and limited available computational power.

2.6.1.3. Computational Fluid Dynamics

The field of Computational Fluid Dynamics (CFD) is a supplemental physical field applied to fuel cell models, which accounts for gas flow characteristics and the related effects on the behavior of the cell. It has attracted some attention during the last years [101, 102] but interest has lagged recently due to advances in microstructure modeling.

2.6.2. Grey Box Models

When trying to establish a white box model, the complexity of the system often prohibits a complete physical description. For this reason, empirical or homogenized relations and parameters are added, resulting in so-called 'grey box models'. The denomination 'grey' indicates a mixture of black and white box models. In fact, different shades of grey are distinguished in [93].

Three types of grey box models will be addressed briefly.

2.6.2.1. Equivalent Circuit Models

ECMs are a powerful modeling tool. By using electrical equivalents, or circuit elements, impedance responses are reproduced. In some cases like the Warburg element (see section 2.3.3.5), which represents gas diffusion in the anode substrate in [39], a physical process can directly be assigned to the circuit element. In those cases the ECMs are similar to a white box model. However, in most cases more empirical circuit elements like the RQ circuit (see also section 2.3.3.5) have to be applied, resulting in a physically motivated but still empirical model – and therefore a grey box model. From the parameters of the electrical equivalents valuable information about the characteristics of the modeled processes can be deduced.

Advantages and the capability of ECM will be further discussed in chapter 5.

2.6.2.2. Homogenized Models

The complexity of microstructure models has already been addressed in section 2.6.1.2. One possible way to handle a microstructure model is to simplify the model by homogenizing sensible processes or spatially distributed variables, as it was done in order to model the charging and discharging of LIBs in [103].

Another method is the introduction of effective parameters, if there is no way to determine the exact values, like it was done in [104] for the bulk diffusion coefficient of a SOFC cathode consisting of LSCF.

2.6.2.3. Multiscale Modeling

A grey box model is the result of the multiscale approach presented in [105]. As first step elementary kinetics models are established for basic electrochemical reactions (see section 2.6.1.1). The results are homogenized (see section 2.6.2.2) and inserted into the next scale – the electrode. The results for the electrodes are again homogenized and inserted into the cell scale, and this method continues for stack and system scales. This way macro-models on the basis of elementary kinetic steps are established.

2.6.3. Black Box Models (Behavior Models)

The purpose of black box models is to reproduce the input-output behavior of a system. Traditional examples for black box models are neural networks and FUZZY logic models [106, 107]. More recently, subspace identification has become popular for the identification and modeling of nonlinear systems [108].

Another practicable way to obtain an accurate model is through experimental measurements. In the case of static models, only a static parameter map has to be parameterized for all output variables depending on all relevant input parameters over the relevant area of operating conditions. For input values that do not lie exactly on the measured grid, the output values are obtained via linear or cubic interpolation. For systems with pronounced nonlinearities, cubic interpolation yields more accurate results, whereas linear interpolation is much faster.

If the dynamic behavior is to be modeled, the approach in [109] yields good results for LIBs and is also applicable for SOFCs. With the help of the Distribution of Relaxation Times (DRT) (see section 4.4) a state-space model for the dynamic behavior of the system in the operating point can be obtained. If this is done for various operating points and the results are compiled in a parameter map for every possible state in the state-space model, a dynamic simulation over the whole relevant operating area becomes possible. It was also shown in [109], that the obtained simulation results showed very small deviations from measured values.

3. Measurement Techniques and their Technical Realization

In this thesis the focus is set on the measurement of current and voltage of electrochemical systems and the possible ways to analyze the obtained data. Hence, this chapter will point out static and dynamic standard techniques and special requirements for fuel cells and batteries to measure these two quantities reliably.

This chapter also includes newly developed measurement techniques. The Time Domain Measurement (TDM) in section 3.6.2 is a new approach of calculating an impedance spectrum using time domain data, that was developed at Institut für Werkstoffe der Elektrotechnik (IWE)[1]. The multisine (section 3.5.2) and the Single Frequency Electrochemical Impedance Spectroscopy (SFEIS) techniques (section 3.5.3) are not new to the field of electrochemical system, however, the first results of these techniques applied to fuel cells and batteries are presented in this thesis.

But at first, the test benches and test samples will be introduced in sections 3.1 and 3.2, respectively. The presentation of measurement techniques starts with static measurements in section 3.3 and proceeds to the introduction of Cyclic Voltammetry (CV) and Cyclic Amperometry (CA) in section 3.4. Electrochemical Impedance Spectroscopy (EIS) as the most important measurement technique for this thesis is subject to section 3.5.1. The presentation of TDMs follows in section 3.6.

A brief introduction to temperature measurement is given in section 3.7.1 and into gas flow measurement in section 3.7.2. these sections provide the theoretical basis for all measurement equipment used in this thesis.

3.1. Test Benches

The reproducible measurement of all quantities mentioned in this chapter requires a sophisticated measurement environment. This section describes the test equipment used at

---

[1]The TDM has already been published in [110].

IWE to ensure such an environment and states the setup used for the measurements in this thesis. As this differs strongly in configuration and specifications for Lithium-Ion Batterys (LIBs) and Solid Oxide Fuel Cells (SOFCs), the test benches used are introduced separately.

3.1.1. Test Bench for LIB

The test benches for LIBs are very simple in principle, but are complex in practice because of safety issues. An advantage for LIB test benches is that every LIB is an enclosed system: reactants and products are confined to the system. Temperature is the only external condition applied. For experiments analyzing the thermal behavior, it should be noted, that current collectors are good heat conductors and the terminals have a considerable heat mass.

A furnace is not sufficient to provide and maintain operating temperature $(T = 0 \ldots 60\,°C)$. There are four possibilities to provide the required stable temperature range for LIB samples realized at the IWE[2]:

- **Controlled ambient air.** If no temperature variation is needed for the experiment, the cell can be tested in ambient temperature, if the laboratory environment is adequately controlled.

- **Environmental chamber.** In an environmental or climate chamber, temperature and humidity can be adjusted. They are commonly used to test the resistivity of products and materials against heat, cold, and humidity. Humidity control is not needed for LIB measurements, because the samples are enclosed systems. Nevertheless, these chambers are widely available and are commonly used in LIB research. The chambers used at IWE are a *Weiss WK1 180* and a *Weiss WK1 340/70*.

- **Temperature chamber.** Temperature chambers only provide the option to control temperature and are therefore perfectly suited for LIB testing. The variety of products on the market is still limited, but with the recent boost in LIB research, temperature chambers are becoming more widely available. The model *Vötsch VT4002* is being used at the IWE.

- **Peltier element.** The Peltier effect can be used to change the temperature of a surface. In the corresponding Peltier element only a current has to be adjusted to pump heat from one side of the element to the other. Devices suitable for LIB testing can be designed. However, most of the systems available on the market have a limited temperature range, which is not appropriate for LIB tests. In [111] a custom-made test setup including a Peltier element and a controller was designed.

For tests with large cells $(Q > 10\,\mathrm{Ah})$, the *Weiss WK1 340/70* environmental chamber is equipped with a CO_2 extinguisher. This precaution cannot immediately stop the burning of a cell, but does cool down the system in case of thermal runaway and fire, and allows for a controlled reaction of the Li with the O_2 in the cell.

Smaller cells $(Q < 10\,\mathrm{Ah})$ do not require a extinguisher unit. Instead a safety chamber controls the exhaust gases and liquids that may leak out of a cell after failure. For this purpose, heavy metal boxes with exhaust channels filled with metal foam are used to prevent

---

[2]The experiments with LIBs in this thesis were conducted in controlled ambient air.

eventual flames from exiting the box and damaging the test chamber. The experimental cells used at IWE (see section 3.2.1) have a very low capacity and no additional measures need to be taken to guarantee their safety, once they are assembled.

Battery test benches are often realized as multi-channel test benches. Some reasons for this are:

- The investment cost for additional channels is quite low and several small cells can be placed together in one climate chamber. Cell housings are reusable and low priced compared to other measurement equipment.

- Experiments have long measurement times because of slow processes in LIB.

- Reproducibility in experimental LIB cells is still problematic, particularly, if a reference electrode is applied.

Electrochemical measurements on cells with a capacity $Q < 10\,\mathrm{Ah}$ are commonly conducted with a multichannel potentiostat and cell test system including Frequency Response Analyzers (FRAs), such as the *Solartron 1470E* used in conjunction with 8 *Solartron 1455FRA*. Due to the high Open Circuit Voltage (OCV) of full LIBs cells, the *Solartron 1260* can only be used for symmetrical cells. Additionally, the *Solartron 1287* electrochemical interface or potentiostat is used for full cell tests. For larger cells ($Q > 10\,\mathrm{Ah}$), a *Zahner IM6* together with a power potentiostat[3] is used. If the cell voltage U_op needs to be determined with very good accuracy, an *Agilent 34970A* may be used to log static and dynamic voltage data up to a sample frequency of $f = 1\,\mathrm{Hz}$ in parallel. In section 3.3.1, the voltage measurement is explained in more detail. Special measurement setups for TDMs and Nonlinear Electrochemical Impedance Spectroscopy (NLEIS) are described in sections 3.6.2 and 3.5.5, respectively.

3.1.2. Test Bench for SOFC

The test benches for the measurement of SOFC single cells are available for two different sizes of active electrode areas at IWE, which will be explained in section 3.2.1. The configuration of the test benches is equal. Both setups were used for the measurements conducted on the corresponding single cells in this thesis. Here, a short summary of the components will be given. A detailed description of the test benches can be found in [39, 67, 112, 113].

- **Furnace.** In a properly insulated SOFC stack, the secondary heat produced during operation is enough to maintain the operation temperature. The heat produced by the small test cells is not sufficient and a furnace is needed to maintain operating temperatures $T = 600\ldots800\,°\mathrm{C}$.

- **Ceramic housing and tubes.** Ceramic tubes are used to direct gases from the mass flow controllers (see below) into a ceramic housing. There the gases are directed to the active electrode areas in a co-flow setup. The ceramic housing has to insulate the cell from the outer gas atmosphere and to separate the anode and cathode atmospheres.

---

[3]Model numbers: *Zahner PP201* ($\pm10\,\mathrm{V}$, $20\,\mathrm{A}$) and *Zahner PP241* ($\pm5\,\mathrm{V}$, $40\,\mathrm{A}$)

- **Burning chamber.** In the burning chamber H_2 and O_2 are brought together and H_2 is burned with the help of a Pt catalyzer net to produce steam (H_2O). It is either located inside the furnace or is mounted externally and is heated to avoid water condensation. This allows for arbitrary mixtures of fuel gas and water vapor to be introduced to the feed gases, thereby adjusting p_{H_2O}.

- **Mass flow controller battery.** Mass flow controllers are used to adjust the gas flow rate Q_{gas} to the electrodes. The experiments in this thesis are all in H_2/H_2O operation with air on the cathode side.

- **DC power supply.** The standard device for adjusting constant currents during a measurement is an *Agilent Power Supply*[4]. Other devices used for EIS and TDM are presented in the corresponding sections.

- **Thermocouples.** For temperature measurement three thermocouples are used as a standard: one for controlling the furnace temperature, one on the cathode and anode sides, respectively. These are approximately $d = 2\,mm$ away from the electrodes to ensure they are not introduced into the gas flow field. Thermocouples as devices to measure the temperature are the subject of section 3.7.1.

- **Data acquisition unit.** The digitalization and recording of the measurement values is realized via an *Agilent 34970A* data acquisition control unit. All measured values (voltage, current, gas flows, temperatures) are recorded every $10\,sec$.

- **Control.** All devices in the test bench are controlled via an integrated software tool, that was developed specifically for the SOFC test benches at IWE.

3.2. Test Samples

In this section, the geometries of the tested samples are presented. The applied materials and the corresponding chemical compositions have already been introduced in section 2.5.1. A list of the samples measured in this thesis is provided in section A.

3.2.1. SOFC

Two geometries of planar Anode Supported Cells (ASCs) were tested in thesis. As depicted in figure 3.1, the active cell areas are the main difference: $A_1 = 1\,cm^2$ and $A_{16} = 16\,cm^2$, both on a $A_S = 5 \times 5\,cm^2 = 25\,cm^2$ substrate.

It will be demonstrated in chapter 5, that the A_1 cell can be regarded as a small scale system and the A_{16} cell as a large scale system, because homogeneous operating conditions can be assumed for the A_1 cell in contrast to the A_{16} cell.

3.2.2. LIB

The LIB tested in this thesis is a commercial Li-ion pouch bag cell with a capacity of $Q = 120\,mAh$. The setup for the experimental cells used at IWE is described in [27, 114].

---

[4]Model numbers: *6612C* ($I_{max} = 2\,A$), *E3646A* ($I_{max} = 3\,A$), *E3644A* ($I_{max} = 4\,A$).

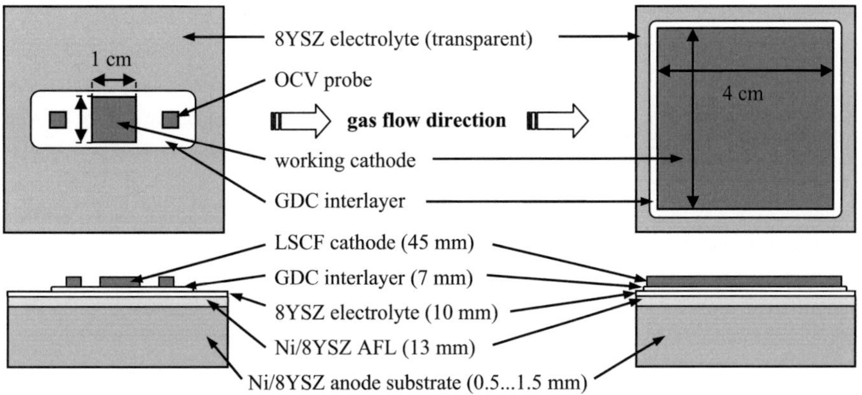

Figure 3.1.: ASC single-cell types used in this thesis: A_1 cell (left) and A_{16} cell (right) in topview (top) and side view (bottom, not to scale).

3.3. Static Measurement of Current and Voltage

In this section the static voltage, current and conductivity measurements are introduced. The measured static voltage is always denominated as U_{op} and the static current as I. As this section is about the measured current in the measurement device, the usage of the current density J will unnecessarily complicate the information provided here. For additional information see also section 2.3.

3.3.1. Voltage

The voltage U_{op} is used to specify the cell voltage. The 4-wire setup used for all voltage measurements in this thesis, allows for U_{op} to equal the voltage measured by the measure-

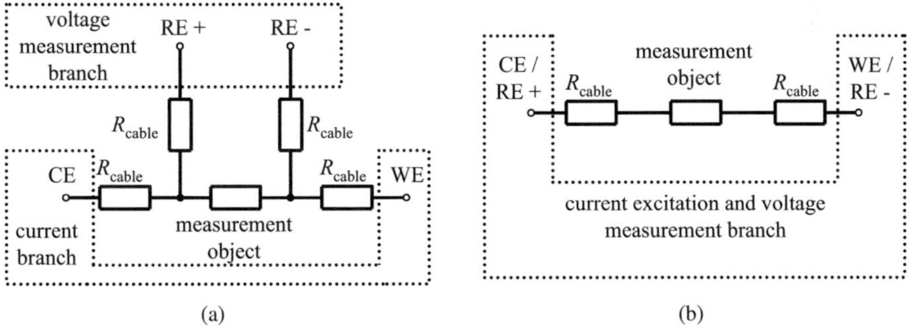

Figure 3.2.: Scheme of (a) a 4-wire measurement setup and (b) a 2-wire setup. In the current excitation branch, the current is adjusted and the voltage is measured at the end of the voltage measurement branch. The denomination of the connections is chosen corresponding to the standard in electrochemical measurements: CE (counter electrode), WE (working electrode), RE+ (reference electrode high), and RE- (reference electrode low).

ment device, U_{meas}. It is defined as the different in potentials, φ at the electrodes used for the voltage measurement,

$$U_{\mathrm{meas}} = \varphi_{\mathrm{high}} - \varphi_{\mathrm{low}} \tag{3.1}$$

A 4-wire setup uses different connections for providing an offset current through the sample and for measuring the voltage of the sample, as can be seen in figure 3.2(a). This way no current flows through the voltage measurement branch, because a good voltage measurement device is characterized by a very high inner resistance. If a very low current in the voltage measurement branch can be assumed, no voltage drop in the connections from the sample to the voltage measurement device distorts the voltage of the sample U_{op}. In an optimized 4-wire setup, the voltage connections are connected near the measurement object, as shown in figure 3.2(a).

A 2-wire setup, as it is depicted in figure 3.2(b), is suitable for an accurate voltage measurement at OCV, because no current is flowing and no voltage drop will distort the voltage measurement. Also for some conductivity and impedance measurements with high resistances, a 2-wire setup is sufficient.

For an accurate voltage measurement, the measurement device must provide a suitable measurement range. The expected voltage to be measured should be about 80% of the maximum voltage of the measurement range. Then it is guaranteed, that the measured values do not exceed the measurement range in presence of measurement noise and that the resolution of the measurement device is used to full capacity.

The resolution of the Analag Digital (AD) converter is given in units of 'bit'. Typical values reach from 8 to 24 bit. This means, that the measurement range is divided into $2^8 = 256$ to $2^{24} = 16.8 \cdot 10^6$ intervals. The importance of a small increment becomes very important for CV and CA measurements, as will be shown in section 3.4.

Basic and advanced circuits for signal adjustment are presented and explained in detail in [25]. Those relevant for electrochemical measurement are summarized in [1]. The static voltages in this thesis were all measured with a *Agilent 34970A* data acquisition control unit. The resolutions and measurement ranges of the measurement devices are difficult to compare, since every manufacturer tends to specify different properties. However, the corresponding webpages provide the required information for the interested reader.

3.3.2. Current

The static currents in this thesis were adjusted with *Agilent Power Supplies* (see also section 3.1). They were measured either by the power supply itself, or via a voltage measurement over a shunt, with the *Agilent 34970A Data Acquisition Control Unit*.

The conversion of the analog current signal into a digital value cannot be done directly as is done for the voltage measurement, because Digital Analag (DA) converters work with voltages. In the following sections, two methods to convert the current into a voltage are presented.

3.3.2.1. Shunt

A shunt is a resistor with accurately known resistance [115]. It is connected in series with the other components in the electrical circuit, in which the current is to be measured. The

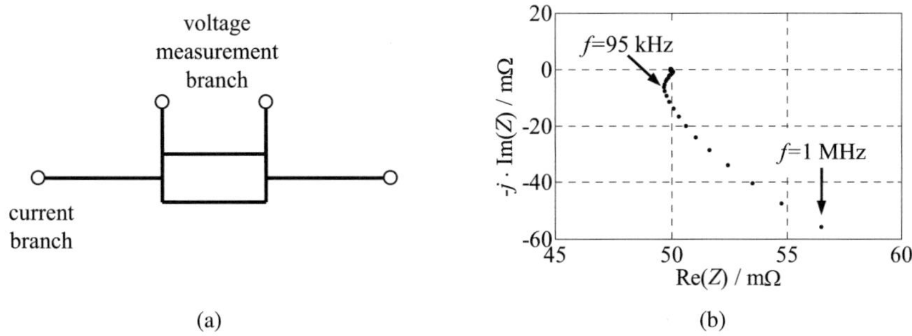

Figure 3.3.: (a) Symbol of a shunt and (b) impedance spectrum of the $50\,\mathrm{m\Omega}$ shunt used at IWE.

two additional connectors of the shunt are used to directly measure the voltage across the shunt in a 4-wire setup. The resistance of the shunt must be small in order to minimize any effect on the characteristics of the electric circuit. The resistance of the shunt must be known accurately and should be temperature stable. This way, the measured voltage drop on the shunt can be converted to a current, using Ohm's law,

$$I = \frac{U_{\mathrm{meas}}}{R_{\mathrm{shunt}}}. \tag{3.2}$$

This is an acceptable method for Direct Current (DC) currents. At high frequencies, shunts show inductive behavior like all resistors, so that the measured voltage is not in phase with the current through the shunt anymore. This can be corrected by an a priori characterization of the shunt for a wide frequency range. The complex impedance, Z_{shunt}, of the shunt is then used to calculate the corresponding current instead of the resistance R_{shunt}. Equation (3.2) changes to

$$i(\omega) = \frac{u_{\mathrm{meas}}(\omega)}{Z_{\mathrm{shunt}}(\omega)}. \tag{3.3}$$

Special high frequency shunts are used for impedance measurements. The type used for the test benches at IWE was measured to evaluate the qualification for a usage at high frequencies, as can be seen in figure 3.3(b). In this figure, the rising inductivity for higher frequencies is clearly visible. Temperature and load stability was tested by applying a constant current of up to $I = 2\,\mathrm{A}$ during the impedance measurement.

3.3.2.2. Magnetic Sensors

Another method to measure the current is to use the magnetic field, that is generated around an electric conductor when a current flows inside it. The physical effect that is used to determine the current is the hall effect [24]. For the current measurement a hall sensor is used. The technique is well-established and a large variety of sensors is commercially available. The advantage to this technique over using shunts is the galvanic isolation of the measurement signal and the current through the sample.

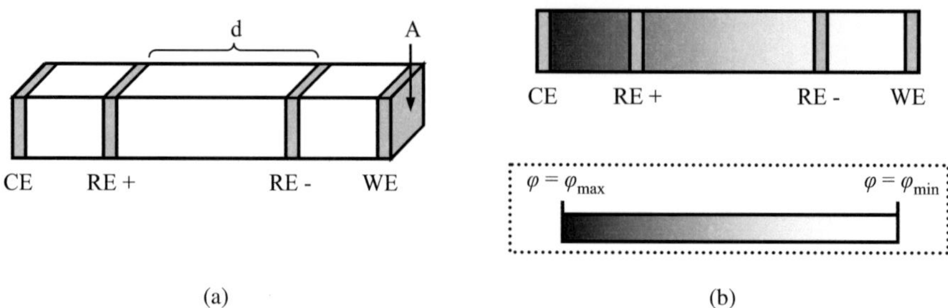

Figure 3.4.: (a) Schematic of a conductivity measurement setup utilizing a 4-probe setup and (b) schematic of an ideal potential distribution in a sample during conductivity measurement with the setup shown in figure 3.4(a).

Other magnetic sensor types exist, such as the *LEM ITZ series*, where a secondary current signal compensates the current inside the sensor induced by the magnetic field of the current to measure I_{meas}. This secondary current is an exact copy of I_{meas}. This sensor principle enables high frequency measurements up to 800 kHz. The market leader for both types of sensors is *LEM*. The major disadvantages of these techniques are the complexity of the sensor and the high price compared to a simple shunt.

3.3.3. Conductivity

The electronic and ionic conductivities of the materials are of interest for a complete characterization of electrochemical systems. Although conductivity measurements are not the main topic of this thesis, some remarks shall be stated here.

For conductivity measurements, a current is adjusted through the sample under test and the voltage drop over the sample is measured. If an electronic conductivity is to be measured, the resistance of the sample determines if a 2-wire or a 4-wire setup should be used, as already explained in section 3.3.1. If contact resistances between the connections and the sample are not negligible, a 4-probe setup must be used, as shown in figure 3.4(a). In this setup two reference electrodes, RE+ and RE-, are used for the voltage measurement. This means, there is no current flowing through them and the contact resistances have no influence on the measurement.

The same is true for the measurement of the ionic conductivity. Here electrodes are needed, which provide the charge transfer from electronic charge to ionic charge, because the measurement device can only provide and measure electronic current. All electrodes used for this purpose are subject to losses in form of resistances as long as a current is flowing through them. This is not the case in the 4-probe setup described above. The voltage drop over the length spanned by the two reference electrodes, RE+ and RE-, is not affected by the electrode resistances, as indicated in figure 3.4(b). Due to the fact that the current in an unbranched electrical circuit is always equal, the current through the sample is the same as the current adjusted by the current source through the electrodes CE and

WE. The conductivity can be calculated as:

$$\sigma = \frac{I}{U_{\mathrm{meas}}} \frac{d}{A}, \quad \text{with} \quad U_{\mathrm{meas}} = \varphi(\mathrm{RE+}) - \varphi(\mathrm{RE-}). \tag{3.4}$$

For conductivity measurements, a combination of a power supply and a data acquisition control unit as described in sections 3.3.1 and 3.3.2 is suitable.

3.4. Cyclic Voltammetry and Cyclic Amperometry

There is another class of methods, that will be included in the overview given in this thesis. These measurements work with very small constant current excitations, I_{CC}, or voltage sweeps with very slow scan rates, s_{CV}, and can therefore be denominated as quasi-stationary measurement techniques or close-to-equilibrium measurements [116]. Quasi-stationarity or close-to-equilibrium state is commonly defined by a maximum deviation of the voltage from the equilibrium state, i. e. the OCV.

The introduction given in this section is not intended to explain CV and CA in full detail. It will only be pointed out in which way these techniques are suitable and how they can be applied to experiments on LIBs for the following purposes:

- Test the chemical stability of a system.

- Identify the insertion potentials of a system.

They are particularly sensible for secondary galvanic cells such as LIBs, because these show a capacitive behavior and the OCV depends on the State of Charge (SOC).

It is difficult to provide a full list of measurement techniques, that belong to the class discussed here, because there is a very large number of methods, that are defined in the literature. However, some measurement techniques are equivalent, even though different names have been used to specify them. Four basic and two relatively new measurement techniques will be introduced with the denominations, that constitute the most frequently used terms in the literature:

- **Cyclic Voltammetry (CV):** CV is a measurement technique applying linear voltage sweeps in both positive and negative direction. In order to visualize the obtained result, the measured current is plotted over the adjusted voltage, as shown in figure 3.5(b). A peak in the CV diagram signifies that a certain electrochemical process is triggered by passing the corresponding voltage, like it is the case for the insertion of Li ions into an anode or cathode material [117]. A peak shift between positive and negative sweep is a measure for the reversibility of the electrochemical process. The slower the scan rate the more pronounced and separated these peaks are. In [117] a very slow scan rate of $s_{\mathrm{CV}} = 5\,\mu\mathrm{V/sec}$ is reported. If the current approaches zero, for the adjusted voltage reaches U_{end} or U_{start} for the second time at t_1, the electrochemical system is stable within the voltage range $U = U_{\mathrm{start}} \ldots U_{\mathrm{end}}$. A decomposition of the electrochemical system is indicated by a significant rise in current towards the decomposition voltage[5]. At this voltage, at least one compo-

---

[5]The decomposition voltage is the voltage until a material is stable. Some materials as YSZ (see section 2.5.2.2) are decomposed into cations and anions above its decomposition voltage which is strongly dependent on oxygen partial pressure and temperature.

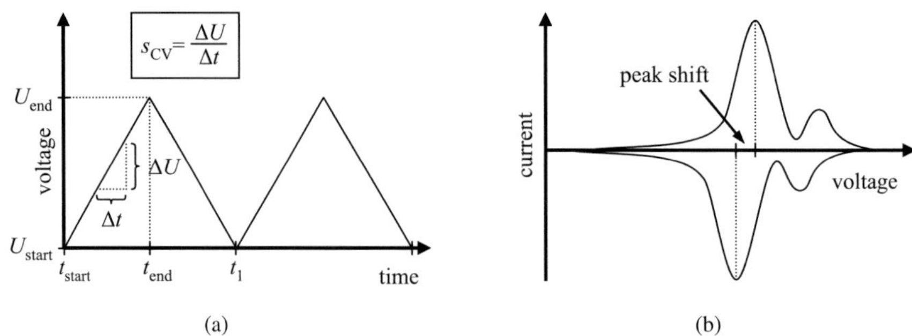

Figure 3.5.: (a) Scheme of the voltage excitation signal of a LSV and CV measurement (LSV: $t = t_{start} \ldots t_{end}$, CV: $t = t_{start} \ldots t_1 \ldots$) and (b) scheme of a CV diagram.

nent is decomposing in positively and negatively charged species, that wander to the opposite electrodes and thereby provide an additional current.

- **Linear Sweep Voltammetry (LSV):** A Linear Sweep Voltammetry (LSV) consists of only one voltage sweep in either the positive or negative direction. It can be said that a CV experiment consists of at least two consecutive LSV experiments with opposite direction. Conducting two LSV experiments with opposite directions provides the possibility to introduce a waiting time during the sweeps in order to let the tested sample fully equilibrate. LSV (or Constant Current Measurement (CCM)) is the experiment, on which Incremental Capacity Analysis (ICA) and Differential Voltage Analysis (DVA) are based, even though experiments with constant current are favored in this thesis, as explained in the next item.

- **Cyclic Amperometry (CA):** CA is comparable to CV, only that the excitation signal is a constant current as shown in figure 3.6(a) and the course of the voltage is measured as depicted in figure 3.6(b). This does not yield an explicit diagram like CV, but has advantages concerning the technical realization of the experiment. Like CV, the direction of the excitation signal is switched at a specific voltage level. Additionally, the current density is limited in this experiment, so that the heat emitted by the electrochemical reaction, which shows a linear dependency on the current density in first approximation (see section 2.4.2), is also limited, facilitating the thermal management of the sample during the experiment. These advantages are also utilized for relatively new techniques such as ICA and DVA, which will be introduced below. The voltage over time curve obtained from the LSV experiment can easily be transformed into a CV diagram by basic mathematical operations, as shown in [118].

- **Constant Current Measurement (CCM):** For the sake of completeness, the CCM is introduced as applying a constant current for a specified time or until a specific voltage level is reached. It is the experiment, on which ICA and DVA are based.

- **Incremental Capacity Analysis (ICA):** This new technique has caught remarkable attention recently and for this, it is addressed here briefly. It is based on LSV or CA,

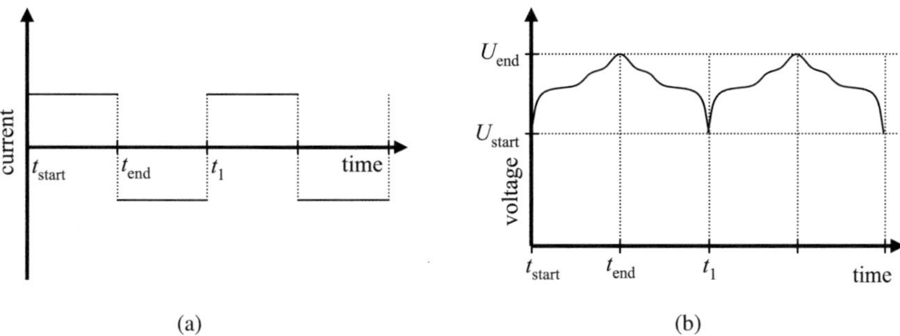

Figure 3.6.: (a) Scheme of the current excitation signal of a CCM and CA measurement (CCM: $t = t_{\text{start}} \ldots t_{\text{end}}$, CA: $t = t_{\text{start}} \ldots t_1 \ldots$) and (b) scheme of a CA diagram.

but gains valuable information about the system by another display type: namely by plotting the incremental capacity $IC = \Delta C / \Delta U$ over the voltage, as described in [116].

- **Differential Voltage Analysis (DVA):** DVA uses LSV or CA measurement techniques as well, but the authors of [119] suggest to plot $-Q_0 \cdot dV/dQ$ over the SOC in order to obtain a characteristic display, where the intercalation processes can be better separated than in a CV analysis.

The two latter techniques have been examined in a recent diploma thesis at IWE [118].

There is a number of devices suitable for the measurement of CV and CA data and these measurement techniques are part of the software packages delivered together with the *Solartron 1470* and the *Zahner IM6* and can easily be designed for a *Novocontrol Alpha A*. Due to the high resolution available, a combination of DC power supply and data acquisition units as described in section 3.3.3 coupled with an adequate control software is recommended.

As discussed in [120], one problem of the CV measurement is the limited resolution of the potentiostat. This is particularly notable for very slow scan rates. The adjusted voltage does not follow a linear sweep, but has a stepped course, where the resolution of the potentiostat becomes obvious, as shown in figure 3.7(a). The obtained current signal is also not smooth. It is rather a series of step responses, as can be observed in the response signal in figure 3.7(b), and it is not clear which points to choose from this signal for data analysis. Fortunately, deviations are very small and good results can be obtained with either minimum, maximum or average of one step. Only the integration of the current that has flown during a voltage sweep from yields significant deviation, if minimum, maximum or average of one step are taken into account. However, a CV measurement has not the intention to determine the capacity of an electrochemical system and this is also not recommended in this thesis. Better data is obtained by the CCM method. Here, only a constant current has to be applied for a certain time. Apart from the measurement noise that superposes the constant current in figure 3.7(d), no inaccuracy in the excitation signal is expected. The voltage response in figure 3.7(c) indeed shows

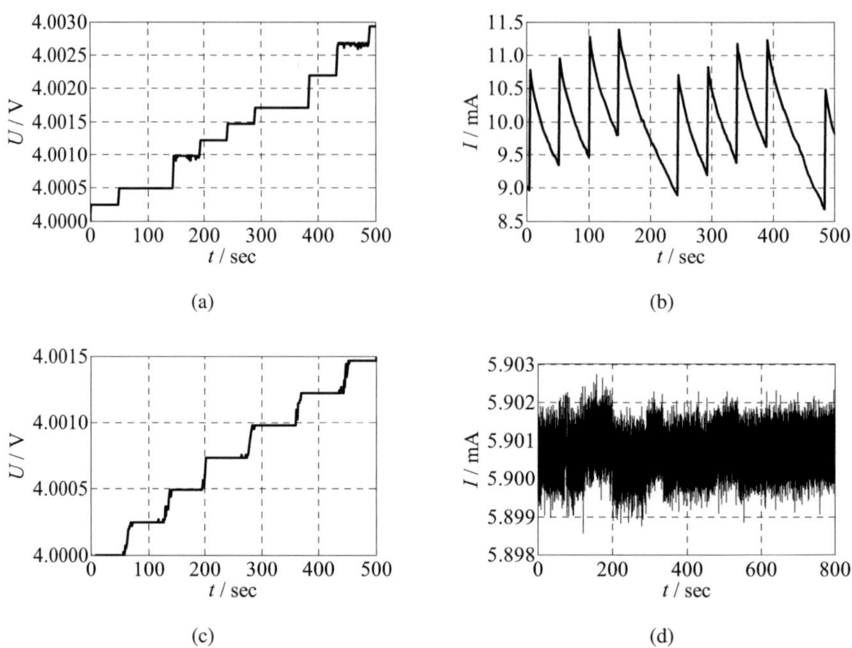

Figure 3.7.: (a) Voltage excitation signal for a CV measurement with a *Solartron 1470*, (b) corresponding current response signal for the voltage excitation shown in figure 3.7(a) with the *Solartron 1470*, (c) corresponding voltage response signal for the current excitation shown in figure 3.7(d) with the *Solartron 1470*, and (d) current excitation signal for a CCM measurement with a *Solartron 1470*. Measurements were conducted with sample L.2.

a similar stepped response as in figure 3.7(a), but in this case it is only a quantization problem of the voltage measurement and behavior of the system under test is smooth. Adequate filters also yield smooth measurement results [118].

3.5. Impedance Techniques in the Frequency Domain

EIS is the most common technique to measure the impedance of an electrochemical system. In the first part of this section, EIS is introduced including practical information about its application. In the subsequent sections, modifications to standard EIS are presented. These comprise the multisine technique (a.k.a. Dynamic Electrochemical Impedance Spectroscopy (DEIS), section 3.5.2), SFEIS (section 3.5.3) and NLEIS (section 3.5.5). Two examples for SFEIS are also provided, demonstrating its capability to analyze specific characteristics of SOFCs. These examples are the temperature measurement via SFEIS in section 3.5.3.1 and the determination of the dynamic change in cathode resistance upon gas variation in section 3.5.3.2.

3.5.1. Electrochemical Impedance Spectroscopy (EIS)

The dynamics of electrochemical energy converters can be measured by EIS. In recent years EIS has been established as one of the most important non destructive characterization methods for electrochemical systems such as batteries or fuel cells. It is applied to model systems as well as to commercially available consumer products. It can be used for fundamental research to analyze the dynamics of basic electrochemical processes as well as for parameter identification of a diagnosis systems for commercial applications. Recently, the history of EIS has been reflected on in [121].

The measured physical quantity is the complex and frequency dependent resistance, $Z(\omega)$, obtained as the so-called impedance spectrum. A valid impedance spectrum can only be obtained if the criteria of causality, linearity, stability and finiteness are met by the measured system [122] (see also section 4.3). Information gained from an impedance spectrum comprises the static behavior of a system at very low frequencies to judge the performance, as well as the dynamic behavior of the system by mapping the polarization processes with different relaxation times as a non-parametrical process model. The analysis of such an impedance spectrum is not restricted to a predefined evaluation method or model assumptions. This offers the possibility to use a variety of specifically capable evaluation methods for analysis, as will be shown in chapter 4.

The measurement itself consists of a series of consecutive measurements at different frequencies. These are commonly logarithmically spaced and defined by the highest frequency f_{low}, the lowest frequency f_{low}, and the (equally spaced) Points Per Decade (PPD). At each frequency point, a sinusoidal perturbation signal is used to cause the electrochemical system to perform a harmonic oscillation around the operating point. Both perturbation signal and system response are recorded over a number of periods. The impedance for the analyzed frequency can be determined via orthogonal correlation with a strictly analogue circuit [36]. More recently developed measurement devices record the signals digitally and calculate the impedance via fourier transform:

$$Z(\omega) = \frac{\hat{U}(\omega)}{\hat{I}(\omega)} \cdot e^{j(\theta_{U(\omega)} - \theta_{I(\omega)})}. \tag{3.5}$$

All quantities on the right side of equation (3.5) are obtained by Discrete Fourier Transformation (DFT). If $F_{a_m}(\omega)$ is defined as the DFT of the measured sequence $a_m(\omega)$, $m = 1, \ldots, M$ with M sample points, then:

$$\hat{U}(\omega) = |F_{u_m(\omega)}(\omega)|, \tag{3.6}$$

$$\hat{I}(\omega) = |F_{i_m(\omega)}(\omega)|, \tag{3.7}$$

$$\theta_{U(\omega)} = \arg F_{u_m(\omega)}(\omega), \tag{3.8}$$

$$\theta_{I(\omega)} = \arg F_{i_m(\omega)}(\omega). \tag{3.9}$$

The obtained values for $Z(\omega)$ are then compiled to the so-called impedance spectrum as explained in section 2.3.3.1. In the following sections, some practical advice is given that should be considered in order to obtain high quality impedance spectra.

3.5.1.1. Type of Perturbation Signal

There are three commonly used ways to provide the sinusoidal perturbation of the system under test for EIS:

- **Galvanostatic:** In this case the perturbation is realized by an alternating current with the amplitude and frequency as parameters to choose. This method is especially appropriate for small impedance values. When the impedance further decreases during the measurement, the peak current is limited by definition.

- **Potentiostatic:** This mode is to be chosen for high impedances, because the optimum perturbation amplitude – mostly $\hat{U} = 5\ldots12\,\mathrm{mV}$ – is guaranteed. For high impedances, there is no danger of a unintentionally high current through the system.

- **Pseudogalvanostatic:** Here, the desired voltage amplitude is chosen for the perturbation, while the measurement device adjusts the corresponding current to reach the desired voltage amplitude. This way, it is ensured, that the voltage amplitude is equal for all frequencies.

3.5.1.2. Linearity

EIS has to fulfil the criterion of linearity. For a the sinusoidal perturbation and response signals for EIS, this means that those signals have to be harmonic oscillations. Linearity of a sinusoidal signal with the frequency f is confirmed if the corresponding fourier transforation shows only one peak at exactly f and is zero otherwise. This is most critical for the system response, but also comprises the perturbation signal.

Linearity of the Perturbation Signal

A good potentiostat or galvanostat has to provide a linear perturbation signal. Problems may occur, if the frequency or the amplitude is so high that the potentiostat cannot provide an adequate signal. The main reason for this is the design of the amplifier stage of the potentiostat. For high frequencies, fast transistors are needed. They are realized by thin n- and p-type doped layers. As the thickness of these layers decreases, the sensitivity increases and limits the maximum current through such a transistor [123]. This is why linearity of the perturbation signal has to be checked particularly for high frequencies and large perturbation amplitudes.

Linearity of the Response Signal

The preconditions for a linear system response – given a linear perturbation signal – are complex and depend on the system under test. As a rule the voltage perturbation should not exceed $5\,\mathrm{mV}$ per electron involved in the electrochemical reaction.

In practice, the output signal should be checked throughout the whole frequency range of the measurement. As this is not possible for most FRAs except the *Zahner IM6* without further measurement equipment, another possibility is proposed: to vary the perturbation amplitude starting with a small amplitude and successively increasing it until a significant systematic change in the impedance is observable. This change is caused by a nonlinear polarization process or by temperature effects due to a too high perturbation amplitude.

It is possible, that for a given voltage perturbation, the system response is linear for high and low frequencies but nonlinear for mid-range frequencies. A possible explanation is a nonlinear polarization process which does not contribute to the response signal to cause a visible nonlinear behavior at high frequencies, because it is not excited. At mid-range frequencies, the nonlinear process contributes significantly to the response signal, so that the nonlinearity is detectable. At low frequencies, all polarization processes contribute to the response signal. This means all processes are excited, but this also means that the perturbation signal is partitioned over all processes, so the contribution of the nonlinear behavior of one of the processes to the overall response is smaller than for mid-range frequencies. Hence, the nonlinearity has a smaller fraction of the response signal than at medium frequencies.

Calculation of Optimum Perturbation Amplitude via Test Measurement

As was shown in section 3.5.1.1, a galvanostatic excitation should be chosen for a small impedance. Nevertheless, the maximum amplitude is always given as a voltage. The optimum excitation amplitude can be determined via a test measurement with a small amplitude at a very low frequency. For this test measurement the result may be noisy due to the (too) small amplitude, however, the DC impedance Z_{DC} can be determined with an sufficient accuracy for the calculation of the appropriate current perturbation amplitude. It can be calculated via

$$\hat{I} = \frac{\hat{U}_d}{Z_{DC}}, \tag{3.10}$$

with $\hat{U}_d$ being the desired voltage perturbation amplitude, and Z_{DC} the impedance at a very low frequency, typically $f \approx 200\,\text{mHz}$ for SOFCs. For LIBs the test measurement should be conducted at the lowest measurement frequency, because $|Z(\omega)|$ rises with decreasing frequency and the required frequency range strongly depends on the purpose of the measurement and the given time.

It should be mentioned here, that different amplitudes are possible for different frequency ranges within the overall frequency range. If different amplitudes are chosen, the frequency ranges should overlap to monitor a possible deviation of measured impedances due to changes in the measurement range applied at the FRA.

3.5.1.3. Signal to Noise Ratio

As it is the case for all measurement methods and their practical realization, a certain level of noise in the data cannot be avoided. EIS has the advantage that the perturbation signal can be concentrated on one observation frequency. Therefore it has a better signal to noise ratio than the multisine technique (see section 3.5.2) and all time domain techniques presented in this thesis (see section 3.6). However, it is important to note that this advantage is correlated with the longest measurement time.

3.5.1.4. Perturbation Amplitude

A trade-off between signal to noise ratio and linearity of the signals is necessary to define an appropriate perturbation amplitude. It should be kept in mind, that different measurement devices define the perturbation amplitude differently. Definitions include:

- $\hat{U}$, the amplitude[6], the factor with which the sine function is multiplied,

- U_{RMS}, the root mean square of one half wave, in the case of a sine function $U_{\mathrm{RMS}} = \frac{1}{\sqrt{2}} \cdot \hat{U}$,

- U_{pp}, the peak to peak amplitude, $U_{\mathrm{pp}} = 2 \cdot \hat{U}$.

SOFC

EIS measurements prior to this thesis at IWE confirmed that A_1 SOFC single cells show a good linear behavior over the whole frequency range $f = 30\,\mathrm{mHz} \ldots 1\,\mathrm{MHz}$, and a perturbation amplitude up to $\hat{U} = 12\,\mathrm{mV}$ can be chosen for the impedance measurement. This is done for all EIS measurements shown in this thesis.

For the A_{16} cell, minor nonlinearities in the current response signal for a voltage perturbation with the amplitude $\hat{U} = 10\,\mathrm{mV}$ occurred at mid-range frequencies, $f = 500\,\mathrm{Hz} \ldots 2\,\mathrm{kHz}$. Therefore, a voltage amplitude of $\hat{U} = 5\,\mathrm{mV}$ for the A_{16} cell was used in this thesis.

It should be noted, that the numbers given in this section have only been tried and checked for the cell types used in this thesis. They may constitute a recommended starting point for EIS measurements on comparable cells. However, linearity should be checked for all new material systems and geometries.

LIB

No general recommendation for the perturbation amplitude can be given for LIBs here, however, $\hat{U} = 5\,\mathrm{mV}$ is a common choice. The test benches for LIBs show less measurement noise compared to SOFCs, because there are fewer external sources of noise such as gas supply and a high temperature heat source. Therefore, accurate impedance measurements are possible with smaller perturbation amplitudes. Common perturbation amplitudes are in the range of $\hat{U} = 1 \ldots 5\,\mathrm{mV}$.

3.5.1.5. Impedance Measurement Setup

The measurement setup for a low impedance measurement differs from a setup for a high impedance measurement. *Low* and *high* are to be considered in relation to the wiring of the tested sample. The resistance of the wiring and the contacts is not negligibly small. For fuel cells and batteries, furnaces or climate chambers are needed to control the temperature. Hence, in order to connect the FRA with the tested sample, the wires usually have a length of $l \approx 1 \ldots 2\,\mathrm{m}$. Together with the contact resistances at the terminals, a resistance of $R \approx 1\,\Omega$ is a good estimate for resistance of the wiring.

Low Impedance

In this work, an impedance $Z(\omega)$ is defined as low impedance, if $Z_{\mathrm{DC}} \leq 1\,\mathrm{k\Omega}$. In this regime, the systematic error with a 2-wire setup becomes higher than $0.1\,\%$, assuming $R = 1\,\Omega$ as resistance for the wiring. Therefore, a 4-wire setup is recommended for measurement of low impedance. The excitation should be galvanostatic as explained in section 3.5.1.1.

---

[6]This value will be used to determine the perturbation amplitude in this thesis.

High Impedance

For high impedances, $Z_{DC} > 1\,\text{k}\Omega$, the error that emerges from the wiring is not critical and a 2-wire setup can be used with minimal error. For very high impedances, the input impedance Z_i of the FRA should be checked. The relation $Z_{DC} \cdot 1000 \leq Z_i$ should hold, in order to keep the systematic error below $0.1\,\%$.

High Frequency

At high frequencies another effect disturbs the measured impedance. This is the inductivity of the wiring. Every electric conductor has a circular magnetic field around it. If the area between input and output conductor is large, it acts like a conductor loop with an inductivity.

To minimize the inductivity, and therefore the influence of the wiring to the measured impedance, the conductors should be arranged as a twisted pair. In this configuration every small residual area between the conductors is ideally compensated by the next one, that points in the opposite direction.

More details about high frequency impedance measurements can be found in [124]. Information about shielding the signal wires and also be found there or in section 3.5.1.6, where a small survey of FRAs is provided.

3.5.1.6. Frequency Response Analyzers (FRA)

In this section, the FRAs used at IWE are presented together with their important characteristics. The information given here is taken from the specifications published by the manufacturers of the corresponding measurement device.

Solartron 1260

The *Solartron 1260* is without doubt the most prominent FRA and has been produced in the current configuration for more than 30 years. It can handle a large range of magnitudes of impedance values and has a very large frequency range. However, the maximum current excitation signal is limited to $I_{RMS} = 60\,\text{mA}$ and the offset voltage is restricted to $U_{OCV} = \pm 3\,\text{V}$ for low frequencies. At frequencies above $f = 1000\,\text{Hz}$ a high pass filter can reject an offset voltage of up to $U_{OCV} = \pm 46\,\text{V}$. Convenient control is provided either via the provided software *Zplot* or General Purpose Interface Bus (GPIB). The shields of the 'input' (counter electrode) and 'generator output' (working electrode) have to be short-circuited in order to enable an impedance measurement. This should be done as near as possible to the measured sample [124].

For larger currents or higher offset voltages the *Solartron 1260* can be combined with the *Solartron 1287* electrochemical interface. It should be paid attention, because the *Solartron 1287* works with driven shields. This means, that the device tries to reproduce the measurement signal on the shields, in order to minimize capacitive effects. These shields must not be connected to avoid damage.

Zahner IM6

The *Zahner IM6* is also a widely used FRA. It can handle similar impedance value ranges and frequency ranges as the *Solartron 1260*, but can provide a larger current excitation signal of $\hat{I} = 3$ A. Control is provided via the *Thales* software package including a large variety of electrochemical measurement techniques.

Also the *Zahner IM6* works with driven shields, that must not be connected to avoid damage (see above).

The *Zahner IM6* can handle higher voltages up to $U_{\text{OCV}} = \pm 3$ V (up to $U_{\text{OCV}} = \pm 10$ V with the *ubuffer*). Further, there are some potentiostats offered by *Zahner*, that provide larger voltage and current ranges. The models used at IWE are *PP201* and *PP241*. They are very suitable for measurements with very low impedance values.

Novocontrol Alpha AK + POT/GAL 15V 10A

The *Novocontrol Alpha AK* is a mainframe FRA, that is normally used in combination with a potentiostat like the *Novocontrol POT/GAL 15V 10A*. It offers a very large impedance value range and a good frequency range. However, high frequency measurements with offset voltages and low impedance values show some deviations.

The device is controlled via the provided software *WinIMP* or *WinDETA* or via GPIB. Driven shields can be chosen as an option in the measurement software.

Solartron CellTest System

The *Solartron 1470* plus *Solartron 1455A FRA* modules is the so-called Solartron CellTest System. It handles up to 8 channels and has a larger current range than the *Solartron 1455A FRA* ($\hat{I} = 4$ A). The voltage range is asymmetrical ($U_{\text{OCV}} = -3\ldots + 10$ V). It can be stated, that the accuracy is not as good as the accuracy of the *Solartron 1260*. Only the provided connectors can be used, because the system is calibrated for these cables. The *Multistat* software controls all 8 channels conveniently.

3.5.1.7. EIS under Load

Most impedance measurements are conducted at OCV. This means, that the current perturbation (or the corresponding current for the voltage excitation) is oscillating around zero. Nevertheless, for some purposes impedance measurements under load[7] are necessary. Then the perturbation signal is superposed by a constant current. This bares the problem, that the current needed for the constant polarization and the actual perturbation current do not necessarily have the same order of magnitude. For example, to measure the impedance of a A_1 SOFC single cell at $U_{\text{op}} = 0.8$ V, a constant current density of about $J = 1.5$ A/cm^2 is required. This means $I = 1.5$ A. A sensible sinusoidal perturbation would be around $\hat{I} = 40$ mA to achieve a voltage response of about $\hat{U} = 12$ mV. This is a problem of resolution for the device as pointed out in section 3.3.1, if the constant part of

---

[7]This type of experiment is called *impedance under load* here, but is often found as impedance under constant polarization in the literature.

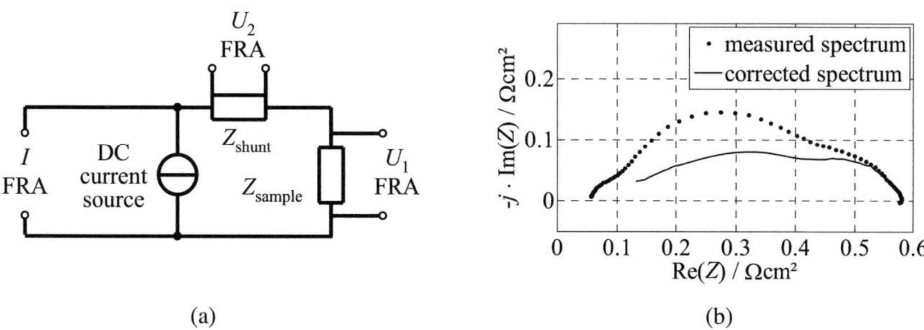

(a)
(b)

Figure 3.8.: (a) Schematic of the electrical circuit that enables consistent EIS under load. (b) Measurement results for sample S.2 for an impedance under load without correction (dotted line, $Z = V1/I$) and with correction (straight line, Z via equation (3.11)).

the excitation signal is considerably larger than the sinusoidal perturbation. The potentiostat has to control a relatively high current while providing an accurate excitation signal that is two orders of magnitude smaller. This is difficult to realize using one potentiostat.

However, both *Zahner IM6* and *Novocontrol Alpha AK + POT/GAL 15V 10A* yield good results for this type of measurement on A_{16} SOFC single cells. Typical values are offset currents of $I \approx 10\,\text{A}$ and a perturbation signal of $\hat{I} \approx 1\,\text{A}$.

Another method to conduct an impedance measurement under load is to use two separate devices, such as a *Solartron 1260* for the impedance measurement and an *Agilent* DC power supply for providing the offset current. The two devices have to be connected with the sample over a special electrical circuit, which is shown in figure 3.8(a). If the two devices are just connected in parallel with the sample, the FRA measures the parallel connection of the sample and the finite impedance of the DC power supply, resulting in the dotted spectrum in figure 3.8(b). With the shunt in the circuit shown in figure 3.8(a), the current in the branch of the sample can be calculated by equation (3.3). The impedance of the sample is calculated via

$$Z(\omega) = \frac{U_1(\omega)}{\frac{U_2(\omega)}{Z_{\text{shunt}}(\omega)}} = \frac{U_1(\omega)}{U_2(\omega)} \cdot Z_{\text{shunt}}(\omega). \tag{3.11}$$

3.5.2. Multisine Technique or Dynamic Electrochemical Impedance Spectroscopy (DEIS)

The multisine technique applies sine-waves of several frequencies simultaneously, as shown in the simple example in figure 3.9.

The response signal is evaluated at the applied frequencies in the excitation signal by Fourier transformation. The obtained coefficients yield the impedance for that particular frequency via equation (3.5). The minimum measurement time depends on the lowest applied frequency, whereas the sampling time depends on the highest applied frequency.

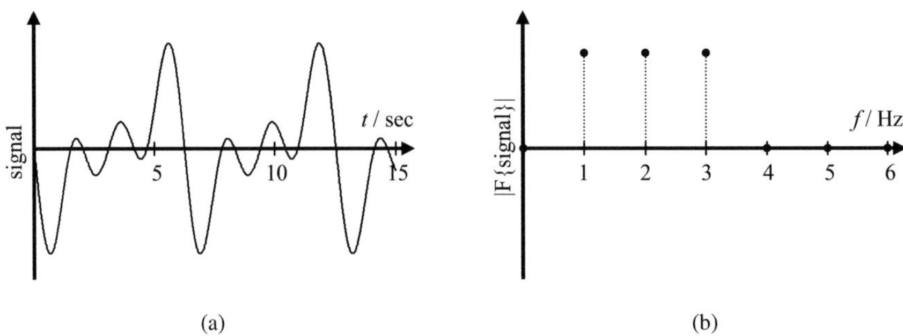

(a)   (b)

Figure 3.9.: (a) Signal consisting of a superposition of three sine functions. (b) The corresponding Fourier transformation to the signal shown in figure 3.9(a).

Therefore, the data volume is a function of the ratio between highest and lowest frequency. In order to avoid high amplitudes in the excitation signal by the superposition of maxima of the basic sine functions, the 'Schroeder multisine excitation signal' [125] and 'odd random phase multisine' [126] are introduced.

Because of the reduction in measurement time and the simultaneous measurement of different frequencies, the multisine technique is used for DEIS in [127]. It is further argued in [128], that the measurement time can be reduced by more than 75% compared to EIS. However, this value depends on the PPD and recorded periods used.

The multisine technique is suitable to identify the evolving impedance of LIBs during charge and discharge as done in [129]. With this technique an impedance spectrum can be pinned to a specific time or SOC with a small alternating excitation signal without changing the SOC of the system, because all frequencies are measured simultaneously.

This measurement technique can be conducted directly with the *Solartron CellTest System*, because the multisine technique is provided as an option in the delivered software. Another possibility is to use a measurement setup, that will be called 'Modular Electrochemical Measurement Setup', consisting of an *Agilent 33250A* arbitrary waveform generator, a *Solartron 1287* potentiostat and an *Agilent MSO6014A* mixed signal oscilloscope. This setup will be explained in detail in section 3.6.1.1.

3.5.3. Single-Frequency EIS (SFEIS)

If a specific characteristic of an electrochemical system can be assessed by the impedance of a single frequency point, this characteristic can be monitored by SFEIS [130, 131]. It is provided as feature in impedance software [132]. But the terminology is not consistent in the literature and there is no comprehensive study giving a general overview of this method. However, it is simply a modification of EIS (or the multisine method limited to only one frequency). To illustrate the capability of this method, two examples for its application to fuel cells will be presented in sections 3.5.3.1 and 3.5.3.2. One example for LIBs including an extension to SFEIS will be announced in section 3.5.4.

3.5.3.1. Temperature Measurement in SOFC

The operating temperature is a crucial parameter for SOFCs. Both performance and long-term stability strongly depend on it, but the real temperature at the reaction zone cannot be measured directly by standard techniques. In this section, a new method for the determination of the operating temperature by means of SFEIS is presented[8].

Motivation

In a test bench under open circuit conditions the temperature can be determined adequately by thermocouples. The housing and the cell are in thermal equilibrium and no heat source or sink is present.

If a constant current density is adjusted in SOFC mode, both a current density through the sample and the electrochemical reaction itself cause heat emission, as explained in section 2.4.2. The temperature within the cell with discernible current densities cannot be assumed as constant. The heat emitted by the cell increases with the current density, as can be seen from equations (2.32) and (2.33).

Therefore, it is highly desirable to determine

- the exact temperature at the interface of anode and electrolyte and in the electrolyte itself, as the majority of heat is emitted there,

- the temperature gradients within the cell to determine the resulting thermal stress, and

- the dynamics with which the temperature is changing within the cell due to load changes, in order to identify pronounced thermocycles.

Concept

As mentioned above, thermocouples are not applicable to measure the temperature within a cell when a current density is applied. Other accessible cell characteristics depend on local cell temperature, like the polarization resistance R_{pol}. However, it also shows a distinct dependence on the current density (see table 5.1 in section 5.3.2), so that the influence of the temperature on the polarization resistance cannot be analyzed separately.

The ohmic resistance R_0 of a cell is defined as the intersection of the impedance curve with the real axis at high frequencies as explained in section 2.3.3.2. It is measured or estimated in a frequency range, in which all electrochemical processes causing an imaginary part in the impedance have already decayed. Hence, the impedance is exclusively ohmic. A change in this ohmic fraction can only be explained by the temperature dependency of the conductivity of the applied materials, that is of Arrhenius type. This is demonstrated in figure 3.10.

It is assumed that the ohmic resistance caused by the electrodes is small compared to the electrolyte and their temperature dependence is also small. The same accounts for the ohmic resistance of the contact meshes [67].

---

[8]The static measurement has already been published in [133] and the dynamic extension was part of a student research project at IWE [134].

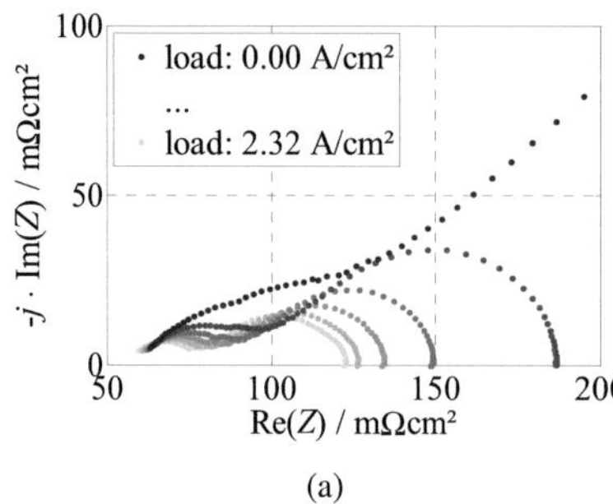
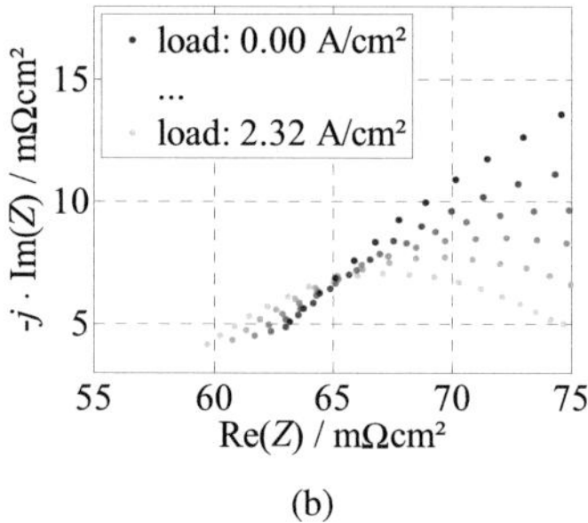

(a)   (b)

Figure 3.10.: (a) Impedance spectra of sample S.3 measured with different loads. (b) Magnification of the high frequent part of the spectra shown in figure 3.10(a).

Hence, the shift of the ohmic resistance of electrolyte is mainly responsible for a change in the measured ohmic resistance, R_0. Unfortunately, it is currently not possible to determine the cell temperature by simply measuring the ohmic fraction of the impedance spectrum of a particular cell, because it cannot be calculated accurately due to remaining uncertainties such as accurate conductivity, thickness, and the residual ohmic resistance of the other cell components. Furthermore, it is difficult to measure R_0 directly because this involves high frequencies ($f > 1\,\mathrm{MHz}$) and low impedance values ($|Z| < 100\,\mathrm{m\Omega}$) at 800°C, which are difficult to achieve. The idea is not to use the measured or fitted value of R_0, but to use the real part of the impedance at only one given frequency, $\mathrm{Re}(Z(f_\mathrm{T}))$, for both a reference curve and a load curve. The frequency must be high enough to avoid influence from the polarization processes and low enough to enable consistent measurement results. It can be seen from the parallel shift of the real part of the impedance, which is shown in figure 3.11(a) for different current densities, that the frequency range between $100\,\mathrm{kHz} \leq f \leq 200\,\mathrm{kHz}$ is suitable. The frequency chosen here was $f_\mathrm{T} = 120\,\mathrm{kHz}$.

With this information, it is possible to calculate the change in temperature by the following method:

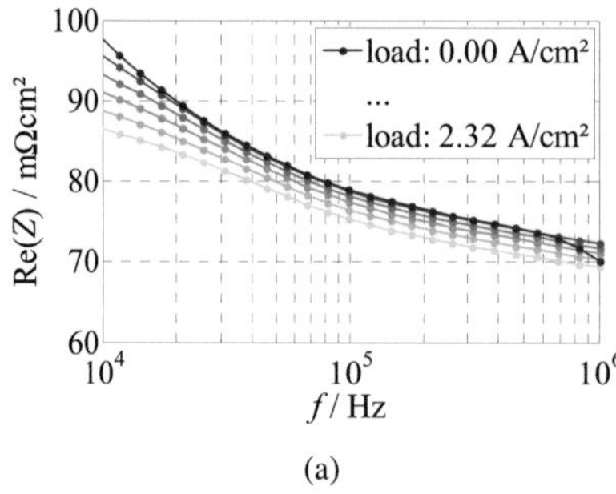
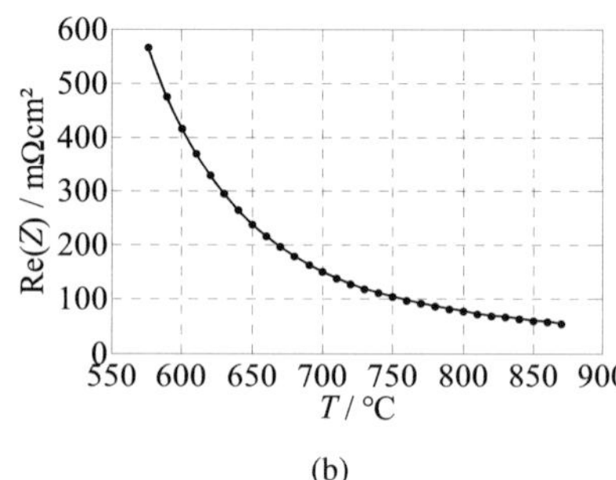

(a)   (b)

Figure 3.11.: (a) Real part of the impedance of sample S.3 shown in figure 3.10(a) at high frequencies. (b) Reference curve: real part of the impedance at $f_\mathrm{T} = 120\,\mathrm{kHz}$ in OCV for different furnace temperatures.

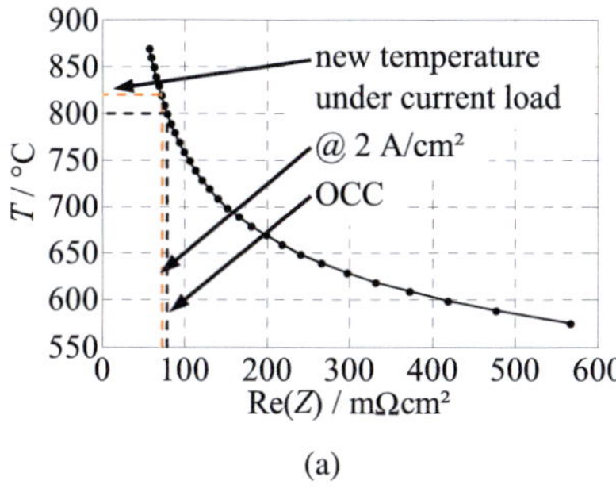

(a)

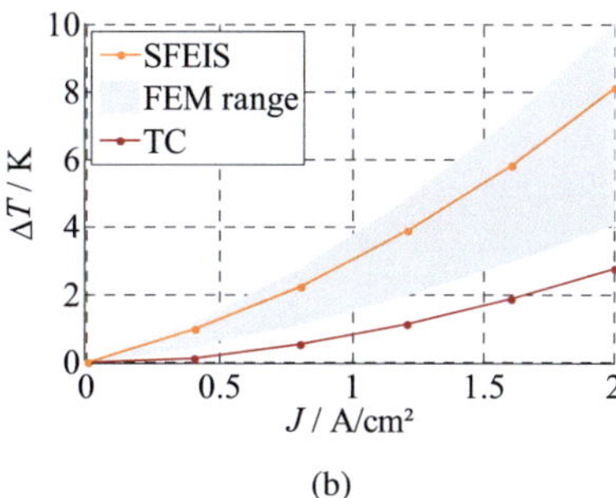

(b)

Figure 3.12.: (a) Inverse of the reference curve shown in figure 3.11(b) depicting the determined cell temperature under load via SFEIS as explained in section 3.5.3.1 for one operating point. (b) Results for the whole operating range in comparison with the temperature measured by thermocouples and simulated by the FEM model as shown in figure 3.13 for sample S.3.

1. Determining $\mathrm{Re}(Z(f_\mathrm{T}))$ for different furnace temperatures at OCV and thus establishing a reference curve for a particular cell without internal heat production. An example for such a reference curve is shown in figure 3.11(b).

2. Measuring $\mathrm{Re}(Z(f_\mathrm{T}))$ under various constant loads, resulting in heat emission by the cell for one fixed furnace temperature (load curve).

The temperature can then be determined by using the inverse function of the reference curve, as demonstrated in figure 3.12(a).

Results

The results are shown in figure 3.12(b), where the increase in temperature is plotted over the applied current density. For comparison, the measured values of the thermocouples are shown in figure 3.12(b) plus the results of a simple Finite Elements Method (FEM) model, where the active area of the A_1 cell in the ceramic housing was modeled as heat source. Details about the used values can be found in [133]. The modeling results are shown in figure 3.13.

Application to SOEC

As mentioned in section 2.4.2 the electrochemical reaction in Solid Oxide Electrolysis Cell (SOEC) mode is endothermic. However, at higher current densities, losses have a larger effect than the thermodynamics. If equation (2.32) is evaluated with data from the static SOFC model presented in [13, 104] (see also section 5.3.2), the heat emission shown in figure 3.14(a) is obtained.

With the SFEIS method derived in this section, the temperature of the cell was determined for both SOFC and SOEC mode. The result is shown in figure 3.14(b) together with the values measured by thermocouples.

A characteristic current density in SOEC mode is the thermoneutral current density, J_tn. This is the current density, at which the losses produce as much heat as is consumed by

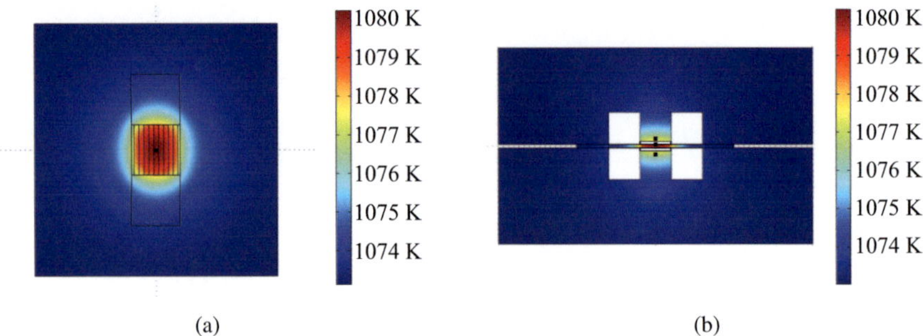

(a)   (b)

Figure 3.13.: Simulation results of the simple FEM model shown in (a) top view and as (b) cross section through the working cathode.

the endothermic electrochemical reaction. It can be seen, that $J_{\text{tn}} = -2.2\,\text{A/cm}^2$ fits well for the calculation shown in figure 3.14(a) and the SFEIS measurement shown in figure 3.14(b). Both methods yield a value of $J_{\text{tn}} = -2.2\,\text{A/cm}^2$. It can also be seen that the overall heat emission is much larger in SOFC mode than in SOEC mode as expected due to the exothermic or endothermic reaction occurring, respectively. It is also confirmed that it is not appropriate to measure cell temperature under load using thermocouples.

Dynamic Change in Temperature (SOFC)

One benefit of SFEIS is its quickness. There is no thermal mass like for a thermocouple, that has to equilibrate with the ambient temperature. With SFEIS, the temperature of the analyzed material itself is measured. For conducting one SFEIS temperature measurement for a SOFC, only a few microseconds are required, because the observed frequency is very high ($f > 100\,\text{kHz}$). That enables a dynamic measurement of the cell temperature upon a change in current density. Sample time is not limited by the measurement itself but the measurement device and the control software. The fastest sample time could be

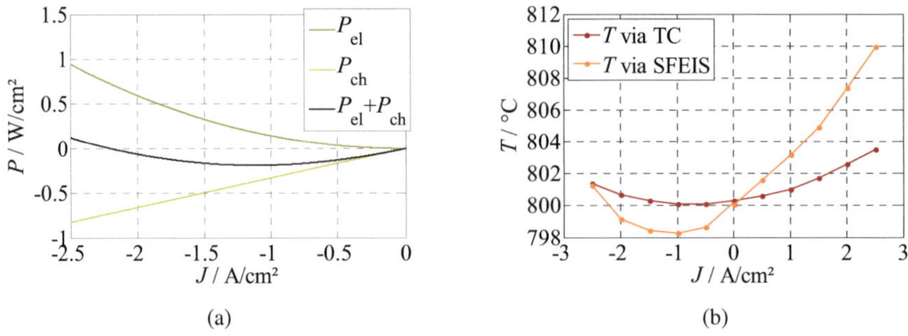

(a)   (b)

Figure 3.14.: (a) Results of the evaluation of equation (2.32) with model data from [13, 104] for SOEC mode. (b) Obtained temperature curve for both SOFC and SOEC operation compared with the values measured externally via a thermocouple during the corresponding measurement with sample S.4.

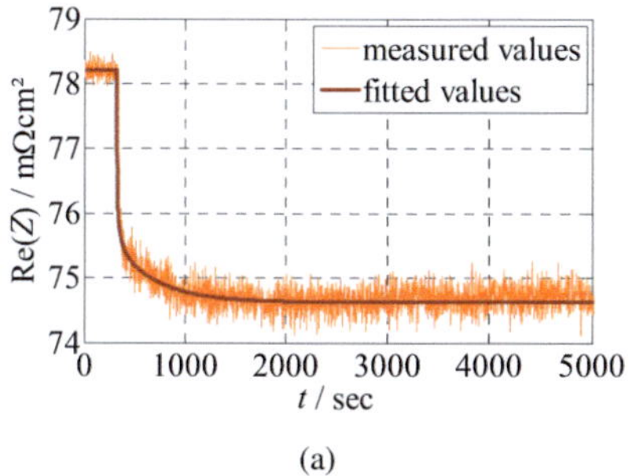
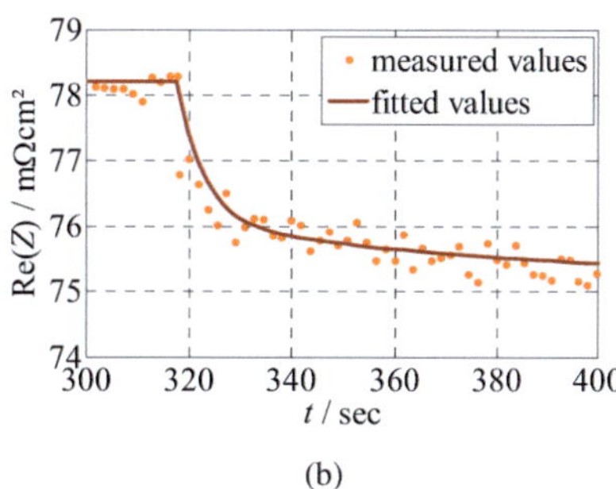

Figure 3.15.: (a) Values for $Z(f_T)$ measured with sample S.3 during a current step plus fitted course. (b) Magnification of figure 3.15(a) at the application of the current step.

achieved by the software *Zplot* and was $T_{\mathrm{sample}} \approx 1.8\,\mathrm{sec}$. The course of $\mathrm{Re}(Z(f_T))$ is shown in figure 3.15 for a current density step from $J = 0\,\mathrm{A/cm^2}$ to $J = 2\,\mathrm{A/cm^2}$ at $t = 317\,\mathrm{sec}$. By fitting the obtained values as shown in this figure, three time constants could be determined, $\tau_1 \approx 5\,\mathrm{sec}$, $\tau_2 \approx 40\,\mathrm{sec}$ and $\tau_3 \approx 500\,\mathrm{sec}$. These were attributed to the thermal behavior of electrolyte, substrate and ceramic housing, respectively.

3.5.3.2. Dynamic Change of Cathode Resistance upon Gas Variation in SOFC

In [10] the dynamic change of the polarization resistance upon a change in the cathode gas was investigated. A symmetrical Lanthanum Strontium Cobaltate (LSC) cathode was operated in synthetic air. Then, 0.037 vol.-% CO_2 was added to the gas. The impedance of the cathode thereupon exhibits a shift, that is too fast to be identified by full impedance spectra. Hence, a SFEIS measurement was started before the gas change and continued over 12 h. After $t = 10\,\mathrm{h}$ the gas change was reversed, and after $t \approx 11\,\mathrm{h}$ the measured resistance had relaxed to the initial value, as shown in figure 3.16(a).

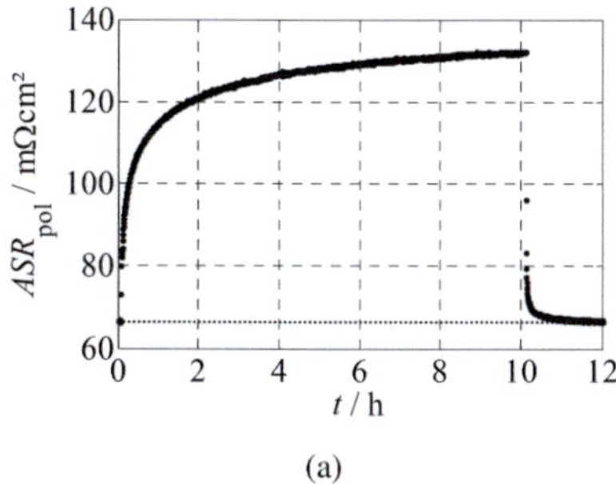
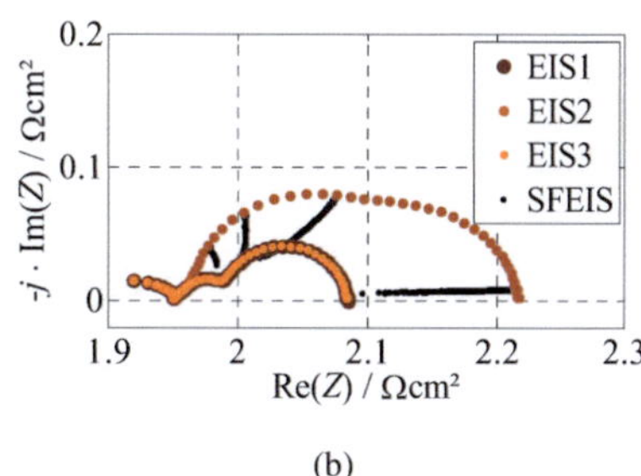

Figure 3.16.: Measurement results showing the dynamic change of the cathode resistance upon gas variation with SFEIS with sample S.5. Figures reproduced from [10].

The SFEIS measurement was conducted with 5 frequencies, which were scanned in rotation ($f_1 = 121\,\text{kHz}$, $f_2 = 562\,\text{Hz}$, $f_3 = 261\,\text{Hz}$, $f_4 = 82\,\text{Hz}$, $f_5 = 1\,\text{Hz}$) as indicated in figure 3.16(b). The frequencies f_1 and f_5 were chosen in order to cover the frequency range of all known polarization processes ($R_{\text{pol}} = \text{Re}(Z(f_1)) - \text{Re}(Z(f_5))$). The frequencies f_2 to f_4 were chosen to monitor the individual loss processes. However, it is difficult to monitor the development of the resistance of individual loss processes by fixed frequencies because the characteristic time constant τ is changing, as can be seen from equation (2.21). More details about the results of this experiment can be found in [10].

There are several characteristics, that can be deduced from the course of the polarization resistance R_{pol} in figure 3.16(a), that shall be addressed here. The obtained data is $R_{\text{pol}}(t)$, a function of time. This data could be used to identify the long-term behavior, i. e. if there is a saturation point or not. Further, by applying time domain methods that will be presented in section 3.6, the data can reveal, if there is one or more processes involved, what share of the whole change in R_{pol} is to be attributed to them, and what the time constants for these processes are.

3.5.4. Electro-Thermal Impedance Spectroscopy (ETIS)

The temperature measurement via SFEIS for SOFCs was already presented in section 3.5.3.1. In addition to that, heat generation in the system can also be triggered by expanding this method. This means the amplitude of a single-frequency has to be modulated periodically to high values. If the heat generation is chosen to be sinusoidal, a sinusoidal temperature response can be determined by the system response, the actual SFEIS measurement. If the excitation is repeated successively at different frequencies, the resulting method is Electro-Thermal Impedance Spectroscopy (ETIS). An introduction, experiments, the corresponding results, and a discussion of this method are published in [50].

3.5.5. Nonlinear EIS

If a harmonic excitation signal for EIS is chosen large enough to provoke the system response to contain higher harmonics, the analysis of them is called NLEIS. As a rule of thumb the excitation signal should be ten times higher than for linear EIS. An introduction is given in [135]. More recent results are shown in [136].

The results obtained from NLEIS are difficult to analyze. No standard analysis method is yet available in the literature. But this does not negate the fact that NLEIS yields more information about the tested sample than EIS. This is demonstrated in figure 3.17(a). A static measurement yields the performance of an electrochemical system, and the individual processes can be identified by measuring and analyzing the unique dynamic features of the involved processes. This is traditionally done by dynamic measurement techniques like EIS. However, EIS only accounts for the linear response of a process that is not necessarily linear. Amplitude and phase of the harmonics in the system response carry additional information, that can give further hints towards an identification of the physical processes in an electrochemical system. This is done in [136], for example.

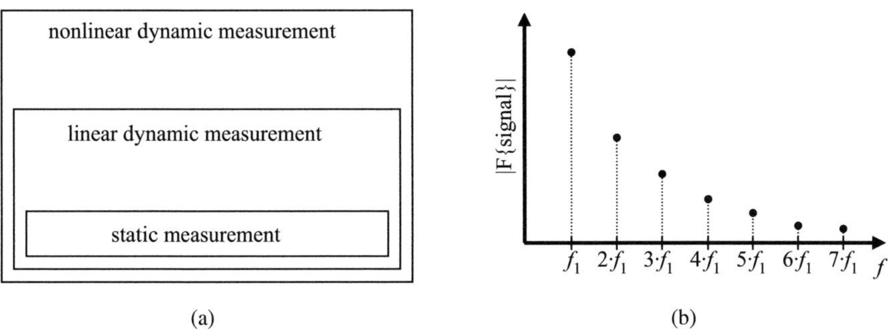

Figure 3.17.: (a) Schematic indicating, that EIS contains more information than static measurement technique, but NLEIS contains even more information about the system under test. (b) Exemplary scheme of the Fourier transformation of a base wave and the corresponding higher harmonics.

Novocontrol Alpha AK + POT/GAL 15V 10A

The *Novocontrol Alpha AK* FRA supports NLEIS. As measurement result the linear impedance plus information about the harmonics are returned. The impedance and the higher harmonics are presented as follows. The impedance at the perturbation frequency ω_1 is obtained via equation (3.5). Correspondingly the higher harmonics for a voltage perturbation are given as:

$$Z_h(\omega) = \frac{\hat{U}(\omega_1)}{\hat{I}(\omega_h)} \cdot e^{j(\theta_{U(\omega_1)} - \theta_{I(\omega_h)})} \quad \text{with} \quad h = 2, 3, \ldots\,^{9}. \tag{3.12}$$

Because of this convention, it is necessary to know the signal $U(\omega)$ in order to reconstruct the signal $I(\omega)$.

Zahner

The *Zahner IM6* provides a visualization of the Fourier coefficients obtained from both perturbation and response signal (also in combination with a *Zahner* power potentiostat, see section 3.5.1.6). If more than the first harmonic can be observed, this indicates, that the corresponding signal is not in harmonic oscillation but is nonlinear. However, only a visualization is provided and the Fourier coefficients are not accessible after the measurement. Hence, only a test of the linearity is possible.

3.6. Time Domain Measurements

Time domain methods are often applied in order to analyze the low frequency behavior of LIBs. For SOFCs, a complete characterization by means of EIS is more common.

This section will introduce general TDM in section 3.6.1. After that, a method to calculate an impedance spectrum out of time domain data will be presented in detail in section 3.6.2.

---

[9] ω_1 indicates the perturbation or base frequency. The first of the higher harmonics or the second harmonic is ω_2.

It is based on [110]. A comparison of measurement times for time domain and frequency domain techniques is provided in section 3.6.3, and at the end of this section a small list of other important time domain methods will be provided in section 3.6.4.

3.6.1. General Time Domain Measurements

Simply put, a TDM is an experiment in which the system under test is subjected to an excitation sequence defined in the time domain and the system response is measured. For the electrochemical measurements in this thesis, current or voltage signals are considered. The corresponding response signal is the voltage or current, respectively.

The chosen excitation sequence has to ensure that the frequency range of interest is adequately excited. Excitation signals are discussed in section 3.6.1.2.

The primary result of a TDM is the course of the response signal over time, from which a corresponding set of time constants can be deduced. These in turn can be used to determine an impedance spectrum, as will be shown in section 3.6.2. Special properties can also be determined by analyzing only specific characteristics of the time domain data, if a specially conditioned excitation signal is applied. Examples here fore include Current Interrupt (CI), Potential Intermittent Titration Technique (PITT), and Galvanostatic Intermittent Titration Technique (GITT), as summarized in section 3.6.4. Fitting the time domain data directly to a pre-assigned model is also a common technique [128].

Integrated solutions for TDMs exist, but the compilation of the input signal is often cumbersome (*Zahner IM6*), limited to a small number of sample points (*Solartron 1470/Multistat*) or restricted to an inherent scheme (*Novocontrol Alpha AK/WinChem*). It is currently not possible to pre-design a sequence in *Matlab* and transfer it to the measurement devices mentioned here. Hence, only very basic excitation signals like step or pulse functions can be realized with these devices conveniently. Another possibility is to use the 'Modular Electrochemical Measurement Setup', which will be introduced in section 3.6.1.1. For the measurements described in section 2.5.2.2 and 3.6.2, a *Novocontrol Alpha-AK* FRA together with a *Novocontrol POTGAL 15V 10A* potentiostat was used.

3.6.1.1. Modular Electrochemical Measurement Setup

This measurement setup was designed and built at IWE for TDM and multisine measurements (see section 3.5.2). It consists of an arbitrary waveform generator, a potentiostat and a digital oscilloscope and offers the possibility to create a signal, excite the measurement object with it, and measure and record both excitation and response signal.

Arbitrary Waveform Generator (Agilent 33250A)

This signal generator has a lot of adjustable pre-defined sequences to design an input signal for the experiment. Furthermore, it is possible to upload self-designed sequences containing up to 64000 individual data points. These sequences can be looped at any chosen frequency within the frequency range of the signal generator ($1\,\mu$Hz$\ldots 80\,$MHz). Note that the frequency range of the signal generator is not the limiting factor in this measurement setup. The frequency range of the potentiostat and the electrical connection are by far more critical. The *Agilent 33250A* has to be connected to a potentiostat and can be controlled by a PC via GPIB using *Matlab* as control software.

Potentiostat (Solartron 1287)

This potentiostat provides the actual input signal to the electrochemical cell and measures voltage and current with high accuracy. But it cannot record the measured values. One great advantage is the fact that the measurement outputs of this potentiostat have the same ground potential and can therefore be recorded simultaneously by a simple multi-channel data acquisition unit. Its frequency range is specified as 1 MHz in potentiostatic mode and 30 kHz in galvanostatic mode, but a decrease in the dynamic range is already noticeable at frequencies higher than 2 kHz. The maximum current is 2 A, the voltage range is 10 V.

It should be paid attention when connecting the potentiostat to the cell, as it works with driven shields, so that a connection of the shields at the test bench can cause serious damage in the potentiostat. The *Solartron 1287* can also be controlled via GPIB using *Matlab*. But as the potentiostat only works as amplifier, only the measurement range and the operation mode have to be adjusted before the experiment. Hence, it is also possible to adjust these parameters with the controls on the front panel.

Oscilloscope (Agilent MSO6014A)

This digital oscilloscope is used for data acquisition. It has four analog inputs that are used to monitor voltage of the electrochemical cell, current through the electrochemical cell, the output signal of the signal generator, and the trigger signal of the signal generator. It should be mentioned that the oscilloscope is a very good device to visualize the measured signals. However, it is not primarily specialized for a fully automated recording of the measured values. To conduct a large number of measurement series, a data acquisition unit in form of a Universal Serial Bus (USB) oscilloscope or a high speed digitizer card are more appropriate.

3.6.1.2. Excitation Signals

There is a large number of excitation signals which can be used for electrochemical measurements in the time domain [1]. In this section, only the very basic signals are discussed.

White Noise

A classical excitation signal in control theory used for correlation analysis is white noise [137]. This is a random signal containing in theory all frequencies ($f = 0, \ldots, \infty$). In practice it is bandlimited, but still contains all frequencies of a certain frequency range with equal amplitude. Therefore, it is suitable to excite an electrochemical system in a large frequency range. However, this technique is not commonly used for electrochemical systems. This is due to the fact that the time constants of electrochemical system are spread over several orders of magnitude and the data volume grows extremely large for an integral identification with white noise (cf. the multisine technique in section 3.5.2). A reduction of data volume as presented in section 3.6.2.1 for a step function is not possible for white noise.

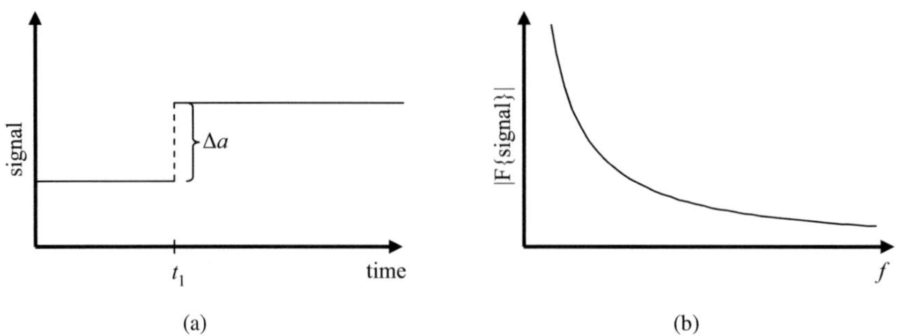

Figure 3.18.: (a) Scheme of a step function as excitation signal and (b) the corresponding Fourier transformation.

Step

Current or voltage steps are commonly used for the identification of electrochemical systems. A step function is depicted together with its Fourier transformation in figure 3.18. If the excitation step provided by the potentiostat is fast enough and the measurement time is long enough, a large frequency range can be analyzed. However, if a large excitation is chosen in order to obtain a pronounced response signal, the operation point before the step and after the step are different. Therefore, a step excitation is only sensible, if the dynamic characteristics of the system before and after the step are alike.

A voltage step as excitation signal is chosen for the TDM for impedance derivation in section 3.6.2. In that section, the advantages and disadvantages of this excitation signal are discussed.

Pulse

A pulse excitation consists of two consecutive steps in opposite direction but with equal amplitude as shown in figure 3.19(a), its Fourier transformation is shown in figure 3.19(b).

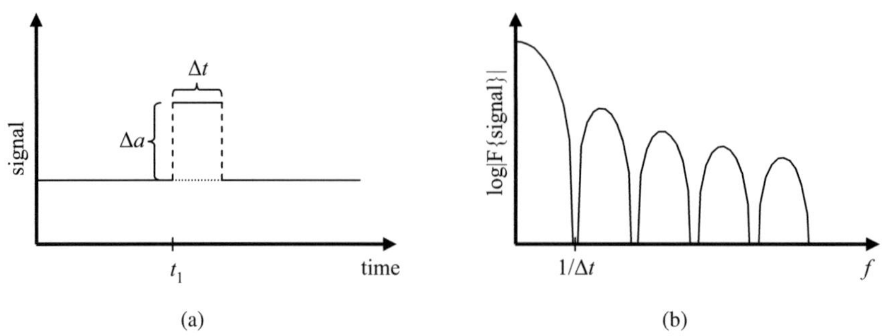

Figure 3.19.: (a) Scheme of a pulse as excitation signal and (b) the corresponding Fourier transformation.

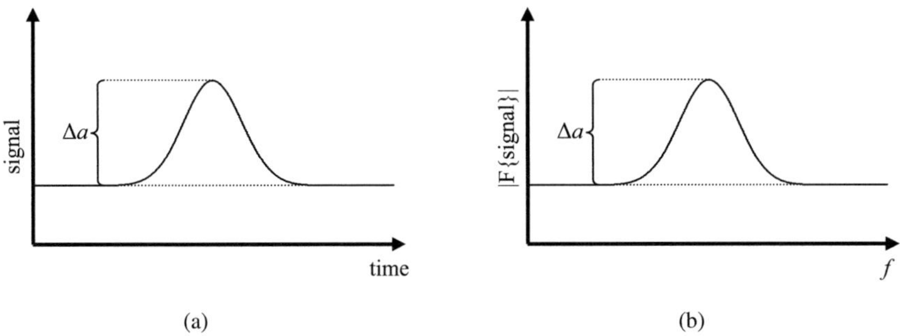

Figure 3.20.: (a) Scheme of a Gauss pulse as excitation signal and (b) the corresponding
Fourier transformation.

The operating point before and after the excitation is the same in the case of a voltage excitation. For a current excitation applied to LIBs the SOC changes by $\Delta Q = \Delta i \cdot \Delta t$.
In order to design an excitation for a given identification experiment, the parameters amplitude and duration have to be optimized. The amplitude of the pulse must not be too
large and the pulse must not be too long to ensure that the tested sample does not heat up
significantly. For LIBs the charge ΔQ of a current pulse should be small enough to obey
the assumption of the same SOC before and after the pulse. Nevertheless, the system has
to be adequately excited, that the system response is resolvable for a long measurement
time. This is necessary to evaluate the system behavior at low frequencies.

Gaussian Pulse

One special excitation signal, that is worth mentioning here, is the so-called Gauss pulse,
which is shown in figure 3.20 together with its Fourier transformation, which is again a
Gauss pulse. It has the best time duration - bandwidth product. This means, it optimizes
the bandwidth of an experiment for a given measurement time as stated in [138]. This will
be further explained in section 3.6.2.2. The mathematical expression for a Gauss pulse is
given in equation (3.17).

This excitation signal is not provided on standard measurement equipment used for this
thesis and no experimental results for the application and evaluation of such an excitation
signal to an electrochemical system could be found in the literature.

3.6.2. Time Domain Measurement for Impedance Derivation

Time domain data can be transformed into an impedance spectrum by using the equation:

$$Z(\omega) = \frac{U(\omega)}{I(\omega)} = \frac{\int_{-\infty}^{+\infty} u(t)e^{-j\omega t}\mathrm{dt}}{\int_{-\infty}^{+\infty} i(t)e^{-j\omega t}\mathrm{dt}}, \tag{3.13}$$

as derived in [139] or by the division of the corresponding Fourier coefficients, as shown
in equation (3.5). For time domain data, these coefficients can be obtained by the direct
application of the Fast Fourier Transformation (FFT) to the excitation and response signal. FFT is a fast algorithm to perform a DFT. This is done in [140, 141], where this

method is called Fourier Transformation Electrochemical Impedance Spectroscopy (FT-EIS). Another possibility is to use the Laplace transformation for this purpose, as was done in [142].

All these techniques are valid for arbitrary excitation signals. As a consequence, they are robust against faulty excitation signals, and need only to be optimized to acquire accurate excitation signals with reasonable measurement times. However, these techniques also entail two major shortcomings: spectral leakage and aliasing. They are originated by the fact that the monitored signals are normally nonzero at the end of the finite monitoring interval and in the finite sampling frequency, respectively. A detailed description of these effects can be found in [137]. Also, the trade-off between information and data volume is an important issue. An approach to overcome these problems will be presented in this section.

The excitation signal for this TDM is a voltage step. The proposed algorithm for the calculation of the impedance spectrum is based on a transformation into the frequency domain as done by Takano in [143]. He used three data acquisition units to realize a variable sampling rate, an exponential window function (realized implicitly by applying the Laplace transformation), and evaluated the time domain data in three sections that were merged.

Significant improvement will be presented in this thesis for the processing of the time domain data. This is achieved by:

- a variable sampling rate realized on one data acquisition unit (see section 3.6.2.1),

- an optimization of the window function (see section 3.6.2.2),

- a transformation of the time domain data using a straight line sequence (see section 3.6.2.3),

- a stepwise evaluation in a large number of sections (see section 3.6.2.4).

3.6.2.1. Data Volume

In order to cover a wide frequency range up to high frequencies with the TDM, a high sampling frequency is required. This reduces the aliasing and results in a better signal to noise ratio for high frequencies. On the other hand, a long measurement time is required in order to evaluate the response at low frequencies (see also section 3.6.2.2).

If these requirements are combined, the resulting data files become extremely large. That is why a step-wise reduction of the sampling rate is proposed here. As already mentioned, a similar approach was used by Takano [143], who used three data acquisition units, that worked on different sampling rates with the same memory, resulting in different observation intervals. The impedance spectrum was calculated in three sections – one section for every data acquisition unit covering a different frequency range.

For the example which will be discussed in section 3.6.2.5 the sampling rate of the measurement device was varied during the experiment according to table 3.1.

interval in sec	sampling rate in Hz
$0\ldots3$	2000
$3\ldots33$	1000
$33\ldots183$	200
$183\ldots783$	50
$783\ldots3783$	20
$3783\ldots33783$	2
$33783\ldots120000$	1

Table 3.1.: Measurement intervals for the example for the TDM in section 3.6.2.5 with corresponding sampling rate.

3.6.2.2. Spectral Leakage

Spectral leakage is the distortion of the Fourier transformation caused by finite measurement time when the signal still has not properly converged to zero. By stopping the measurement, all the information contained in the signal at later times is implicitly assumed to be zero. Mathematically the thereby occurring distortion $W(\omega)$ of a spectrum can be described by folding the undistorted spectrum with the Fourier transformation of a rectangular function of the width T_{meas}, which is given by

$$W(\omega) = 2T_{\mathrm{meas}}\frac{\sin \omega T_{\mathrm{meas}}}{\omega T_{\mathrm{meas}}}. \tag{3.14}$$

As $W(\omega)$ is oscillating and its absolute value is decreasing slowly towards higher frequencies, the distortion can become remarkably large and occurs in a wide frequency range. Spectral leaking can be minimized by multiplying the measured data with an appropriate window function with better distortion properties than the implicitly used rectangular window function. Common window functions in signal processing are Hamming, Blackman [137], or exponential functions ([143], implicitly). For the problem discussed here, the optimal window function is as narrow as possible in frequency domain and thereby – to achieve short measurement times – as fast as possible decreasing to zero in the time domain. The claim for a narrowband window function is according to the interpretation of the folding operation as sliding averaging of the undistorted spectrum, weighted with the spectrum of the window function. According to [144], the property of a function to fulfil both requirements can be described by the so-called time duration - bandwidth product. It gives an estimate for the relation of the width of a function in the time domain, the so-called time period Δ_t, and the width of its spectrum in the frequency domain, the so-called bandwidth Δ_f. It is

$$\Delta_t\Delta_f \geq \frac{1}{4\pi}, \tag{3.15}$$

which is a version of the uncertainty principle. Hence, a small time duration of a function causes a large bandwidth and vice versa. It is also shown in [144], that the function with the smallest possible time duration - bandwidth product, given by

$$\Delta_t\Delta_f = \frac{1}{4\pi}, \tag{3.16}$$

is the so-called Gaussian pulse,

$$w(t) = e^{-bt^2} \quad \circ\!\!-\!\!\bullet \quad W(\omega) = \sqrt{\frac{\pi}{b}}\,e^{-\frac{\omega^2}{4b}}. \tag{3.17}$$

Therefore, for a given measurement time the Gaussian pulse is the optimal window function which provides the least distorted voltage and current spectra.

The parameter b of the Gaussian pulse has to be chosen in order to guarantee a small residual value of the windowed signals at the end of the measurement time to avoid additional spectral leaking by an implicitly used rectangular window function. It could be shown experimentally that

$$b = -\frac{2\pi}{T^2_{\text{meas}}} \tag{3.18}$$

results in a good trade-off between small residual value in time domain and small distortions in the frequency domain [145].

The remaining error rises with decreasing frequencies. Hence, the lowest theoretically possible frequency of $f_{\text{min}} = 1/T_{\text{meas}}$ exhibits a large error. This is demonstrated by a model system consisting of a resistor and a finite space Warburg element in series. In Figure 3.21(a) the real and imaginary parts of the impedance are plotted together with the values calculated from the windowed time domain data of a voltage step. A measurement time of $T_{\text{meas}} = 105\,\text{sec}$ was simulated, corresponding to $f_{\text{min}} = 10^{-5}\,\text{Hz}$. The sampling rate was kept constant at $f = 100\,\text{mHz}$. Due to the results of this simulation, the minimum frequency that can be evaluated with this method is proposed to $f_{\text{min}} = 4/T_{\text{meas}}$. In this case the error is kept lower than 2% (cf. figure 3.21(b)), which is a relatively low value for very low frequencies. This error can be avoided by applying a Gaussian pulse as the excitation signal. This theoretically reduces the error caused by spectral leakage to nearly zero, because no further windowing of the excitation signal would then be necessary. Another advantage of a Gaussian pulse originates from the minimal time duration -

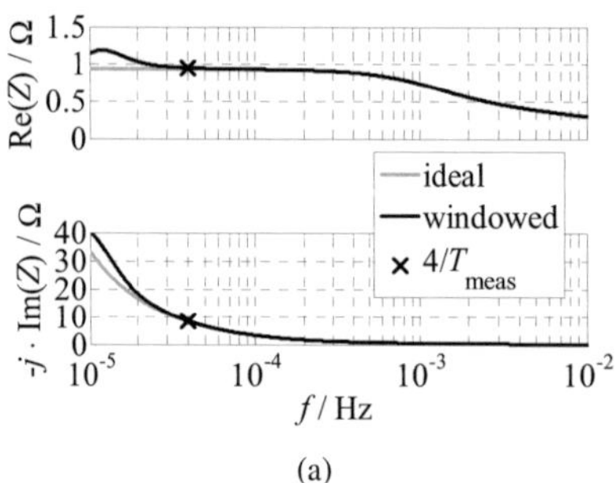

(a)

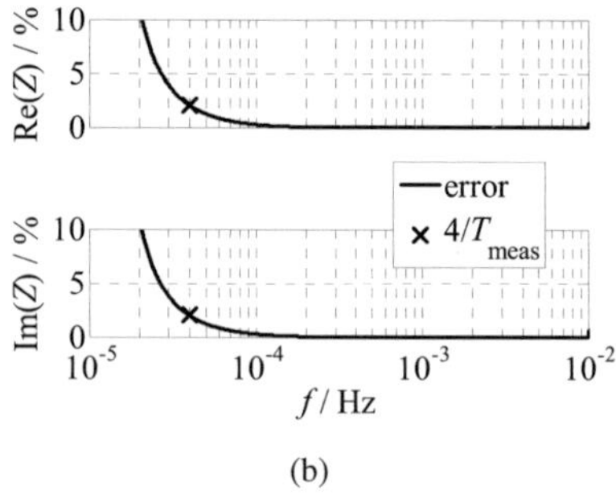

(b)

Figure 3.21.: (a) Real part and imaginary part of impedance of a model system consisting of a finite space Warburg element in series with a resistor for the ideal case (dotted line) and calculated with windowing of time domain data with a Gaussian pulse given by equation (3.17) and the parameter a defined according to equation (3.18) (continuous line), (b) error in the real and imaginary part.

bandwidth product. It shows the best concentration of signal energy in the low frequency range for a given measurement time – an ideal feature for low frequency measurements. A Gaussian pulse as excitation signal was already considered in [138], however, as already stated in section 3.6.1.2, no experimental results for the realization and evaluation of TDMs with this excitation signal could be found in the literature.

3.6.2.3. Aliasing

To minimize aliasing effects during transformation, a piecewise linear discrete transformation algorithm was developed. This algorithm connects the discrete measurement points by straight lines to a straight line sequence and thus directly reconstructs a continuous signal without the need for any a priori assumptions. This circumvents the error originating from the numerical evaluation of the Fourier integral through DFT. This matter was studied in detail in [146].

The sequence is transformed directly by applying an algorithm derived from adequate correspondences of the Fourier and Laplace transformation, as can be found in [139]. This algorithm is based on an algorithm for equally spaced measurement points from [147], but was rewritten for non-equally spaced measurement points as follows.

The time function $\hat{h}(t)$ of a straight line sequence for non-equally spaced measurement points h_n with $\Delta t_n = t_{n+1} - t_n$ is given by

$$\hat{h}(t) = \underbrace{\left(\sigma(t) - \sigma\left(t - \sum_{i=1}^{1}\Delta t_i\right)\right) \cdot \left(h_1 + \frac{h_2 - h_1}{\Delta t_1}t\right) + \dots}_{\text{first straight line sequence}}$$

$$\underbrace{+ \left(\sigma\left(t - \sum_{i=1}^{n-2}\Delta t_i\right) - \sigma\left(t - \sum_{i=1}^{n-1}\Delta t_i\right)\right) \cdot \left(h_{n-1} + \frac{h_n - h_{n-1}}{\Delta t_{n-1}}t\right)}_{\text{last straight line sequence}}, \quad (3.19)$$

where $\sigma(t)$ is the normalized step function. The transformation of this function leads to

$$\hat{H}(\omega) = \frac{1}{j\omega} \cdot \left(h_1 - h_n \cdot e^{-j\omega \cdot \sum_{i=1}^{n-1}\Delta t_i}\right)$$

$$- \frac{1}{\omega^2} \cdot \left(\frac{h_2 - h_1}{\Delta t_1} - \frac{h_n - h_{n-1}}{\Delta t_{n-1}} \cdot e^{-j\omega \cdot \sum_{i=1}^{n-1}\Delta t_i}\right)$$

$$- \frac{1}{\omega^2} \cdot \sum_{k=1}^{n-2}\left(\frac{h_{k+2} - h_{k+1}}{\Delta t_{k+1}} - \frac{h_{k+1} - h_k}{\Delta t_k}\right) \cdot e^{-j\omega \cdot \sum_{i=1}^{n-1}\Delta t_i}. \quad (3.20)$$

This formula allows for the direct calculation of the Fourier integral of the straight line sequence out of measurement points h_n and the corresponding time intervals Δt_n between them. In contrast to a numerical evaluation of data points with the Fourier or Laplace integral, there is no numerical error in this transformation, as it represents the analytical solution of the Fourier transformation of the straight line sequence.

The impedance $Z(\omega)$ is then calculated via equation (3.13), with $U(\omega)$ and $I(\omega)$ being the transformed straight line sequences obtained from equation (3.20) for the measured voltage and current signal, respectively.

3.6.2.4. Evaluation in Sections

The calculation of the impedance spectrum is done in as many sections as possible. The selection of calculation intervals is thereby not restricted to the intervals given in table 3.1. For noise reduction the number of intervals is further increased. For the experiment shown here, 38 calculation intervals $(t_{0,k}, t_{\mathrm{end},k})$ were chosen as follows:

$$t_{0,k} = 0, \quad t_{\mathrm{end},k} = 10^{-2+\frac{k}{5}} \quad \forall k = 0, \dots, 38. \tag{3.21}$$

Similarly, the number of sections Takano uses for the calculation of the impedance spectrum is determined by the number of different sampling frequencies [143]. This has the purpose of reducing aliasing: When the sampling rate of a data signal is reduced during the measurement time, the lowest included sampling rate f_{low} produces a damped repetition of the spectrum exactly at f_{low}, if the entire data set is transformed. According to the Shannon theorem, no frequency higher than half of the sampling rate can be analyzed without aliasing.

Beyond that, further increasing the number of sections has benefits in noise reduction: This can be explained by the Fourier integral. After a certain time T_1, no further information for frequency $f = f(T_1)$ but only noise is added by continuing the integration. This time T_1 cannot be specified a priori. Therefore a large number of intervals was evaluated, so that the noisy high frequent part of each interval can be replaced by the low frequent part of the subsequent interval.

Note that all measured data points in the corresponding interval are used for calculation. The algorithm used for transformation in this work (see section 3.6.2.3) can handle non-equally spaced measurement points. Hence, no down-sampling of the first intervals to the slowest sampling rate within the calculation interval – in order to obtain equally spaced measurement points – is necessary.

The obtained 38 sections of the spectrum are merged by a simple algorithm that verifies the agreement of the first derivation of the absolute value of the impedance. This algorithm also specifies the transition frequency and by that defines the relevant frequency range for each section. It also guarantees a smooth transition by a simple weighting function.

3.6.2.5. Example for an Impedance Spectrum Calculated from Time Domain Data

The ability of the presented technique is demonstrated by experimental data of a commercial Li-ion pouch bag cell[10] with a capacity of $Q = 120\,\mathrm{mAh}$. The starting conditions of 95 % SOC at a constant ambient temperature of 21°C were adjusted and the cell voltage settled at $U_{\mathrm{OCV}} = 4.1882\,\mathrm{V}$ after 3 days, so that all processes in the relevant frequency range could be assumed to have decayed. First an impedance spectrum (EIS 1) was recorded using a *Novocontrol Alpha-AK* FRA together with a *Novocontrol POT/GAL 15V 10A* in a 4-wire-setup. The excitation was chosen to $\hat{U} = 10\,\mathrm{mV}$. The frequency range was $f = 33.4\,\mu\mathrm{Hz}\dots10\,\mathrm{kHz}$. There was a delay of one period before three periods were recorded the for impedance calculation for every frequency point.

The focus for this measurement was set on accuracy. Therefore less frequency points were measured with longer integration time than for a standard cell characterization. This

---

[10]Sample L.2, as listed in section A.

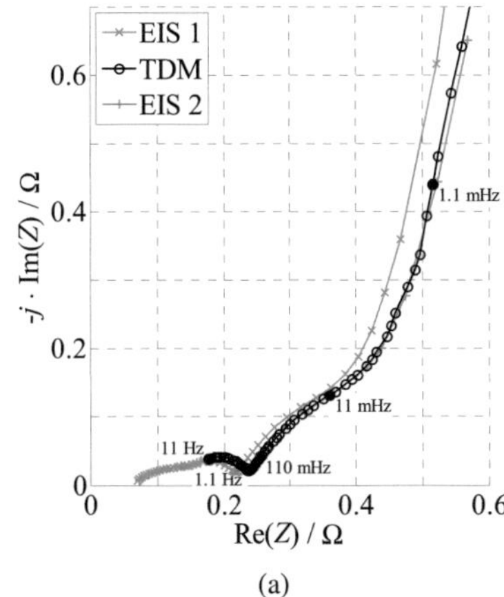 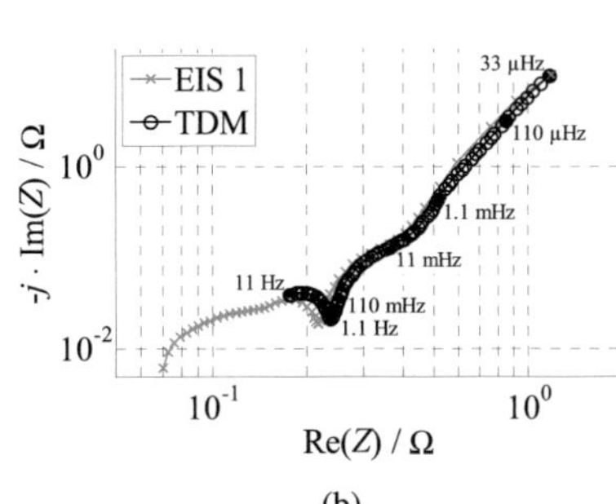

(a)

(b)

Figure 3.22.: Measurement results of the TDM with sample L.2. (a) Nyquist plot of the high frequent part and (b) logarithmically scaled Nyquist plot of the impedance spectra of a commercial LIB obtained by EIS (EIS 1 and EIS 2) and the TDM proposed here; good agreement of EIS 2 and TDM is shown; the accuracy of the measurements is demonstrated by Kramers-Kronig test in figure 3.23.

added up to a measurement time for EIS 1 of approximately 2 days. Afterwards, the cell was again held for 2 days at $U = 4.1882\,\text{V}$. Then, the TDM was started. At $t_{\text{start}} = 0$, a voltage step of $\Delta U = -50\,\text{mV}$ was applied using the same measurement device. Current and voltage were monitored for 1.5 days. The sampling frequency was varied in intervals as shown in table 3.1.

Afterwards, a voltage step of $\Delta U = 50\,\text{mV}$ was applied to restore the starting conditions. When the cell had settled at $U = 4.1882\,\text{V}$ after 3 days, another impedance spectrum (EIS 2) was recorded with the same specifications as EIS 1. To reduce measurement time, EIS 2 was stopped after reaching $f = 680\,\mu\text{Hz}$ after approximately 2 h.

The time domain data was processed by a *Matlab* script as described above. The obtained spectrum covers the frequency range from $f = 33\,\mu\text{Hz} \dots 9.5\,\text{Hz}$ with 15 PPD. EIS 1 covers the frequency range from $f = 33\,\mu\text{Hz} \dots 10\,\text{kHz}$, whereas the lowest frequency measured for EIS 2 is $f = 680\,\mu\text{Hz}$.

A comparison of the data obtained by EIS and TDM is shown in figure 3.22(a) (high frequent part) and figure 3.22(b) (whole spectrum as logarithmic Nyquist plot). It can be observed, that TDM and EIS 2 are in very good agreement, while EIS 1 shows a considerable deviance.

To analyze this fact, the data quality of the three spectra was tested by the Kramers-Kronig test presented in section 4.3.2.2. The residuals obtained by this test are depicted in figure 3.23 and give an account of causality, linearity, stability and finiteness of the tested spectrum.

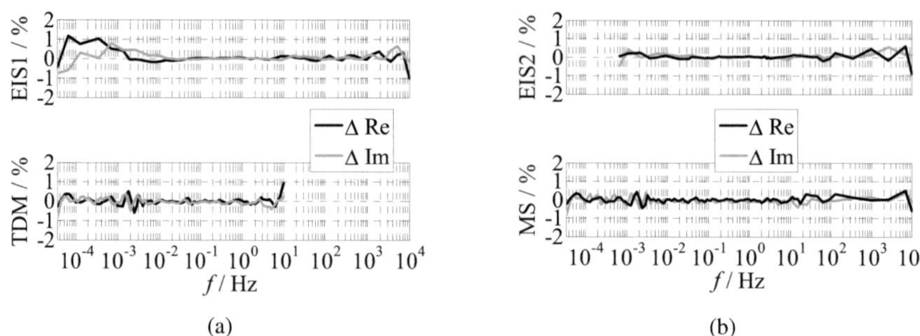

Figure 3.23.: Kramers-Kronig residuals for EIS 1, TDM and EIS 2 shown in figure 3.22, and for the merged spectrum (MS) of TDM and EIS 2.

The following conclusions can be deduced from the results of the Kramers-Kronig test:

- The residuals of EIS 1 show a systematic deviation at low frequencies. Therefore, the tested sample was not stable during the measurement (duration 2 days).

- The residuals of TDM and EIS 2 only show minor noise over the whole frequency range indicating that the criteria of causality, linearity, stability and finiteness were met for these measurements.

The reason for the unstable behavior of EIS 1 during 2 days is yet unclear, but it must be considered that it is linked to the measurement technique. Probably it was caused by a current drift or inaccurate excitation signal, because the temperature was kept constant throughout the whole experiment and the settling time before the measurement was sufficient. During the $120000\,\mathrm{sec}$ (1.5 days) of TDM, the 2 days settling time and the $2\,\mathrm{h}$ of EIS 2 the system was considerably stable. This is demonstrated by the low residuals of both TDM and EIS 2 and the very good accordance of these two measurements. Hence, only by the TDM it was possible to obtain a good impedance spectrum at the lowest observed frequencies.

It should be noted that even though a spectrum calculated via Fourier transformation should always fulfil the Kramers-Kronig test, this test is sensible for the here proposed TDM, because the spectrum is composed of different sections that are the result of the evaluation of different monitoring intervals, that have been merged by a simple weighting algorithm.

It also should be mentioned that in [148] similar results for the impedance spectrum of a cell with comparable material composition were reported.

3.6.2.6. Conclusion

The good accuracy of the TDM was already demonstrated in section 3.6.2.5. Another advantage of the TDM is the high frequency resolution, that can be observed in figure 3.22(b). With this technique 15 frequency points are evaluated per decade as a standard. A doubling of frequency points is also possible causing only a longer calculation time. So a higher frequency resolution does not go along with a longer measurement time, as it is for EIS.

At high frequencies, time domain techniques face the critical issue that the excitation signal has to be adjusted fast with very good accuracy by the potentiostat. This is a more critical task than providing a harmonic oscillation at high frequencies as standard FRAs do. The sampling rate of the data acquisition equipment must also be correspondingly high and accurate at the same time.

Therefore, EIS has advantages at higher frequencies, where it has a better signal-to-noise ratio because the whole signal energy of the excitation signal is concentrated in one frequency. The overall signal energy is limited for all techniques, because it has to fulfil the linearity condition of the test sample. As the measurement time is not critical above 10 Hz, EIS is preferred for frequencies higher than 10 Hz. It is possible to measure the frequency range from 1 MHz to 10 Hz with 15 PPD and at least 10/100 delay/integration periods in less than 2 min.

As the main conclusion, a combination of TDM and EIS is considered as the optimum technique to evaluate a large frequency range of electrochemical systems, resulting in a merged spectrum (MS). The Kramers-Kronig residuals for MS are also shown in figure 3.23. The merger of MS was realized by the same technique as explained in section 3.6.2.4.

The measurement time of the whole measurement adds up to to $T_{\mathrm{meas,TDM}} = 4/f_{\min}$ for the TDM plus approximately $T_{\mathrm{meas,EIS}} = 2\,\mathrm{min}$ for EIS (that is less than 1 % of the total measurement time, if frequencies below 20 mHz shall be examined).

Potential for a further reduction of measurement time of the TDM is given by the application of a Gaussian pulse as excitation signal as described in section 3.6.2.2.

3.6.3. Comparison of the Measurement Times for Time Domain Measurements and Impedance Spectroscopy

The TDM takes up less measurement time when compared to EIS, as is also concluded in [128, 149]. However, the time-savings depend on the measurement parameters chosen for the two methods. Generally, a settling time has to be provided for the system for both measurements, so that all relevant processes in the observed frequency range have decayed.

Beyond that, a delay period is important for an accurate result using EIS. This is demonstrated by the following theoretical consideration. An electrochemical system – in this example a RC circuit ($R = 1\,\Omega, C = 500\,\mathrm{mF}, f_{\mathrm{RC}} = 1/(2\pi RC) = 330\,\mathrm{mHz}$, cf. figure 3.24(a)) in series with an ohmic resistor – is assumed to be in harmonic oscillation. At the time t_1 the frequency of the perturbation signal is switched from frequency $f_1 = \omega_1/(2\pi)$ to $f_2 = \omega_2/(2\pi)$, in order to evaluate the subsequent frequency. Surely the system will not immediately oscillate harmonically with frequency f_2.

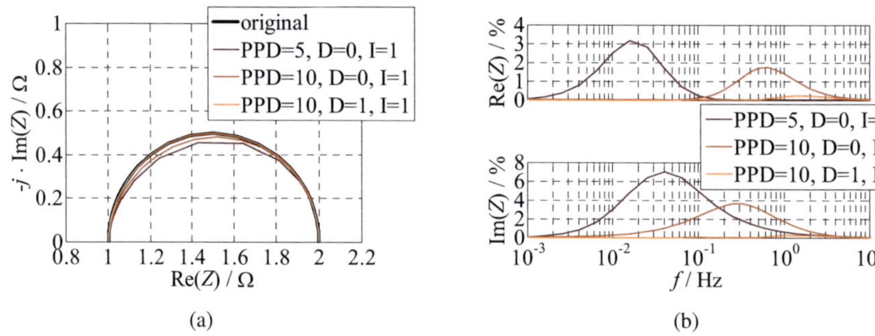

(a)　　　　　　　　　　　　　　　　　　(b)

Figure 3.24.: (a) Nyquist plot showing the comparison of a model system consisting of a RC circuit in series with an ohmic resistor and the results of the simulation of impedance spectroscopy measurements on this model system without noise, applying the corresponding delay periods (D) and integration periods (I). (b) Relative error in the real and imaginary part of the impedance in the simulation of the impedance spectroscopy measurement of the model system shown in (a) resulting from the corresponding measurement parameters applied.

In order to analyze this behavior, the deviation Δi from the harmonic oscillation is calculated and is given by the following deviation term, with $\hat{U}$ as the voltage excitation:

$$\Delta i(t) = \hat{U} e^{-\frac{1}{RC}(t-t_0)} \left(\frac{\omega_1^2 RC^2}{1 + (\omega_1 RC)^2} - \frac{\omega_2 C \cos\left(\frac{\omega_2}{\omega_1}\frac{\pi}{2}\right) + \omega_2^2 RC^2 \sin\left(\frac{\omega_2}{\omega_1}\frac{\pi}{2}\right)}{1 + (\omega_2 RC)^2} \right).$$

(3.22)

As can be seen, the magnitude and the duration of this term depend on

1. the values of R and C;

2. the ratio of f_1 and f_2; therefore, the deviation term grows, if less frequency points per decade are measured;

3. the distance of f_1 and f_2 from the characteristic frequency f_{RC} of the system; the deviation term has its maximum in the proximity of f_{RC}.

These statements are also emphasized by the simulations shown in figure 3.24(a). The error-free impedance response is depicted together with the calculated result of an impedance measurement. In figure 3.24(b) the corresponding errors are plotted. The measurement parameters for these simulations were chosen as follows:

- 5 PPD (points per decade), 0 D (delay periods), and 1 I (integration period);

- 10 PPD, 0 D, and 1 I;

- 10 PPD, 1 D, and 1 I;

without applying measurement noise. It can be observed that the error is significant without delay periods and gets worse when only few points per decade are measured.

type	PPD	T_meas in h	$T_\mathrm{meas,EIS}/T_\mathrm{meas,TDM}$
TDM	15	111	-
EIS	5	147	1.32
EIS	10	266	2.40
EIS	15	386	3.48

Table 3.2.: Theoretical measurement times for EIS with a different number of PPD compared with the time domain method presented in this thesis ($f = 10\,\mu\mathrm{Hz}\dots 10\,\mathrm{Hz}$). One delay period and one integration period was chosen for EIS. Note that this is the minimum time to guarantee a consistent result.

Also, it should be mentioned that a longer integration time reduces the influence of measurement noise. Therefore, a measurement time of two periods (one as delay and one for integration) is seen as the minimum time for each frequency point for EIS. This corresponds to Barsoukov's comparison of the measurement time of his TDM with the multisine technique and EIS in [128]. It is argued in [128] that the TDM presented there requires only 12% of the measurement time compared to EIS. However, it has to be noted that this value is only possible, if the system satisfies the assigned model and the measured data can be extrapolated beyond the monitored measurement time.

With the method proposed in section 3.6.2 – without any model assumptions – the minimum measurement time T_meas for a desired minimum frequency f_min with a step function as excitation signal can be estimated to $T_\mathrm{meas} = 4/f_\mathrm{min}$ as shown in section 3.6.2.2. In table 3.2 this measurement time is compared with that of EIS for different measurement parameters. Compared to EIS the measurement time could be reduced by 71.2% by the TDM for the same frequency density without any a priori assumptions.

3.6.4. Other Time Domain Methods

There is a large number of specific TDMs as already mentioned in section 3.6.1.2. Three of them shall be addressed briefly in this section because they have attracted some attention in recent years.

3.6.4.1. Current Interrupt (CI)

One realization of a TDM is the so-called CI technique. Here fore the (high) current drawn from a system is simply cut off at a certain time and the voltage response is measured. This technique is proposed for the determination of the impedance of low impedance systems such as Proton Exchange Membrane (PEM) fuel cell stacks at high frequencies [150]. It is favored for this purpose to avoid inductive effects presenting a problem for Alternating Current (AC) methods.

CI is also proposed for fast diagnosis of electrochemical cells with large time constants. This is done by directly fitting the time domain data to a pre-assigned Equivalent Circuit Model (ECM) as described for the CI technique in [151]. The point at issue for these methods is that the choice of the ECM is based on assumptions determined before the measurement as discussed in section 3.6.3.

In [152], it is reported how a CI measurement can be conducted through a power electronics control element included in a test bench.

3.6.4.2. Galvanostatic Intermittance Titration Technique

There are two special techniques to determine diffusion coefficients in LIBs in the time domain. For the first one, GITT, a current pulse is applied to the battery and the voltage response is measured.

Due to the current pulse the voltage rises because of the thermodynamic activity of the inserted species [153]. It is shown, how the diffusion coefficient can be determined from the linear slope if the voltage is plotted over $\sqrt{t}$.

3.6.4.3. PITT

The second technique to determine diffusion coefficients in LIBs with an experiment in the time domain is PITT [154].

Here a voltage step is applied that is causing Li ions to adsorb immediately on the electrode surface reassuring chemical and electrical equilibrium. Due to the high concentration gradient within the electrode, the outer compensation current is maintained. The concentration gradient decays over time, so that the current also decays. If the measured current is plotted logarithmically over the time, the diffusion coefficient can be identified from the slope of the obtained diagram, if the exact thickness of the electrode is known.

3.7. Measurement of Further Physical Quantities

The correlation between current and voltage in electrochemical systems is the most important for their characterization. However, the identified model equations mostly depend on other external physical quantities, too, that are interpreted as operating parameters or ambient conditions.

For batteries they are limited to the ambient temperature. The Current/Voltage (C/V) characteristics of fuel cells depend on more operation parameters. These are:

- temperature,
- anodic gas composition (including flow and pressure), and
- cathodic gas composition (including flow and pressure).

In order to obtain reproducible measurement results, these quantities have to be monitored during each experiment, too. For this reason, this section will give a brief introduction into the basics of the measurement of these further physical quantities.

3.7.1. Temperature Measurement

Generally, there are several well-established techniques to measure a temperature. But the applicability of these techniques depends on the ambient conditions and the temperature range of the test setup. If the temperature at the reaction zone of a SOFC shall be measured, several problems emerge. The sensor has to be inert for oxidizing and reducing gas atmospheres. As all these requirements are impossible to meet, the best suiting of the following options has to be chosen.

Thermocouple

Thermocouples work on the principle of the Seebeck effect [155]. Every electronic conductor has a Seebeck coefficient that specifies the thermovoltage of the conductor due to a temperature difference between the two ends of a wire. If two materials with different materials are combined, the corresponding thermovoltage can be measured.

The advantages of this technique are the wide temperature range from absolute zero to the melting point of the materials applied, a variety of material combinations for different sensitivities at different temperature ranges, the inertness if noble metal is used, and the simple measurement setup without the need of a power supply.

Thermocouples of type 'S' are used in the SOFC test benches as a standard.

PTC / NTC

Another method to measure the temperature is to use the effect of the change in resistance of a material due to a temperature change. Materials with Positive Temperature Coefficient (PTC) and Negative Temperature Coefficient (NTC) exist and are applied for the corresponding sensors. All of these have to impress a current into the sensor and measure the voltage drop.

A variety of materials is applicable for different applications. The well-known Pt100/Pt1000 sensor is made of platinum, which has a positive temperature coefficient and is used in many applications.

Pyrometer

The pyrometer is an elegant way to measure the temperature on the surface of a body. It has a wide temperature range and can determine surface temperatures without contacting the measurement object. However, due to high costs and the difficult installation in a test bench, thermocouples are preferred for the temperature measurement in the test benches at IWE.

Impedance Spectroscopy

In section 3.5.3, it was shown how the temperature of a SOFC in operation can be determined by SFEIS.

In the case of the SOFC, the advantage of this type of measurement is that the temperature is determined directly at the electrolyte by its ohmic resistance. The spot, where the temperature is determined, is directly in the thin electrochemically active cell compound and is not affected by the cooling of the gas flow, as it would be if determined at the cell surface. The integration of a passive temperature sensor at this spot is not possible, because of the harsh conditions and the limited space.

The major drawback for this technique is the need of a reference curve and a FRA as described in section 3.5.3.

3.7.2. Gas Flow and Composition Measurement

A few remarks on gas composition measurement and gas flow measurement shall be given here.

3.7.2.1. Gas Composition Measurement

As shown in section 2.3.1, the OCV of a fuel cell depends only on temperature and the gas compositions at the anode and cathode side. This fact can be used to measure the shares of a binary gas composition. For SOFCs in H_2/H_2O operation and the assumption of ambient air on the cathode side, the ratio of H_2 and H_2O can be calculated by the Nernst equation (2.13) for a given operating temperature. This is done via the OCV probes described in section 3.2.1.

More complex gas compositions are measured with gas chromatographs as a standard. Those devices allow the determination of different gases included in the observed gas flow. However, steam has to be removed from the analyzed gas, because the gas chromatographs usually work at ambient temperature and the steam would then condense in the gas chromatograph or in the pipes leading to it. This is why steam cannot be measured as a gas component by gas chromatographs. The interested reader is referred to [48] for a more detailed introduction to the working principle and the interpretation of measured data of gas chromatographs.

3.7.2.2. Gas Flow Measurement

The mass flow controllers in the SOFC test benches also have to measure the gas flow in order to be able to control it adequately. This is done by thermal mass flow meters. The measurement principle works as follows: heat is introduced into the gas flow stream that shall be determined. At a second position downstream the gas flow, they measure how much heat has been dissipated with a temperature sensor.

The heat dissipation can be correlated to the mass flow of the gas, if the heat capacity of the gas is known. This is one reason, why mass flow controllers have to be calibrated for one special gas. The other reason is the chemical compatibility of the applied sealing materials and the gas.

4. Analysis of Measurement Data

In chapter 3 different measurement techniques have been presented. The impedance spectrum is regarded as central result of the techniques, because it can be obtained both by time domain and frequency domain techniques (cf. sections 3.5.1 and 3.6).

In this chapter different possibilities of how to draw performance characteristics, model parameters, general system dynamics parameters and physical properties out of impedance spectra will be shown.

But first of all, it is essential to define the operating point, in which the spectrum is recorded. Therefore, the chapter starts with the analysis of static measurements in section 4.1. After that, the analysis of impedance spectra is discussed in detail. This starts with their visualization in section 4.2, before the Distribution of Relaxation Times (DRT) is introduced as powerful tool for the identification of electrochemical processes in section 4.4. Methods how to verify impedance spectra are discussed in section 4.3. In section 4.5 the Equivalent Circuit Model (ECM) method is presented, as it is conducted as a standard to obtain impedance contributions of different processes to the spectra. This is followed by the presentation of some new approaches consulting the DRT in order to conduct a pre-identification of impedance spectra (section 4.5.2) and to stabilize the fit and to gain a better resolvability of the electrochemical processes in the fit by applying the DRT as additional quality criterion (section 4.5.4)[1]. This is seen as valuable contribution to draw additional information out of impedance data compared to standard techniques such as Complex Nonlinear Least Squares (CNLS) fit, and to lead to a better understanding of electrochemical systems.

4.1. Analysis of Static Measurement Data

Despite the amount of information contained in an impedance spectrum, analyzing the static measurement data should not be omitted, because it is defining the operating point of the electrochemical system. This is information that cannot be drawn from an impedance

---

[1]This method has already been published in [156].

spectrum alone. Therefore, the analysis of the Open Circuit Voltage (OCV) is discussed briefly in section 4.1.1. The performance of an electrochemical system is still most conveniently analyzed via Current/Voltage (C/V) characteristics as explained in section 4.1.2.

4.1.1. OCV

The OCV alone can provide essential information about the system under test. As already stated in section 2.3.1, no distinction is made between the theoretical cell voltage and the terminal voltage under open circuit conditions, U_{OCV}. The difference between them is mostly due to residual electronic conductivity in the electrolyte. When gaseous reaction partners are involved, gas leaks in the measurement setup or in the electrolyte can also result in a lower U_{OCV}. However, these differences are very small for the considered systems and measurement setups.

The residual conductivity is small for Lithium-Ion Battery (LIB) electrolytes as can be deduced from the low self-discharge rate [12]. At technically relevant operating conditions, the Yttria Stabilized Zirconia (YSZ) electrolyte for Solid Oxide Fuel Cells (SOFCs) also shows an electronic conductivity, but one that is several orders of magnitude lower than the ionic conductivity and can therefore be neglected [31].

The further characteristics will be discussed separately for SOFCs and LIBs.

4.1.1.1. SOFC

The OCV for SOFCs is determined by the oxygen partial pressure difference between anode and cathode. As pointed out in section 2.3.1, the theoretical voltage of a fuel cell is given by the Nernst equation (2.12) and (2.13), respectively.

For fuel cells in H_2/H_2O operation, the measured OCV does not correspond well to the voltage calculated by equation (2.13) with respect to the gases supplied to anode and cathode. This can be attributed mostly to minor gas leaks in the test bench. Ambient air leaks into the ceramic housing, increasing the p_{O_2} on the anode side. This results in a decrease of the ratio of $p_{O_2,cathode}$ and $p_{O_2,anode}$ and leads to a decrease in cell voltage given by equation (2.12). The cathode gas is not critical, because mostly ambient air is used here and a gas leak to the outer gas atmosphere has no influence on the measurement. A leaking of anode gas to the cathode side, or vice versa, is not considered here because of the design of the ceramic housing. Nevertheless, this would have the effect of a voltage decrease for the anode as described above, and no effect on the cathode side, as ambient air contains a share of 21% H_2O and a minor leak would not affect this value significantly.

The desired gas atmospheres on anode and cathode side can be adjusted via the measured OCV. The OCV can be measured very accurately as shown in section 3.3.1 and represents reliably the oxygen partial pressure gradient between anode and cathode. The desired gas composition can be adjusted through the O_2 content in the anode gas, which reacts with H_2 to produce H_2O in the burning chamber (see section 3.1.2). When the calculated voltage for the desired gas composition is reached, this gas composition can be assumed as the current operating condition of the tested cell. The largest deviation is expected at very dry conditions, when only hydrogen is supplied to the anode. Typical leakage rates for these conditions are below $0.5\,\%$ for the test benches used at Institut für Werkstoffe der Elektrotechnik (IWE).

4.1.1.2. LIB

For LIBs, the OCV is the key quantity to determine the State of Charge (SOC) of the cell, given it is in steady state. The accuracy hereby strongly depends on the characteristics of the voltage over OCV curve. For the cathode material, $LiFePO_4$ for example, a relatively flat curve is obtained so that an accurate determination of the SOC is more difficult than for the steeper curve of the cathode material Nickel Manganese Cobaltite (NMC).

4.1.2. C/V – Charge Characteristics

The measurement of the C/V characteristics of fuel cells and batteries differs fundamentally because of the finite reservoir of charge carriers in batteries and the infinite reservoir of charge carriers in fuel cells (cf. section 2.1).

4.1.2.1. SOFC

A constant current excitation results in a steady-state operation for fuel cells. The C/V curve of a fuel cell reflects the performance of the cell in the operating point. But the actual measure of the performance is subject to ambiguity. Different physical values can be consulted to express the performance, as pointed out earlier in section 2.3.4. There, the power density and the power loss density were proposed as adequate measures to judge the performance of a fuel cell.

The performance analysis then consists of the definition of an operating point and a C/V measurement at that specific operating point.

The operating point is commonly defined by temperature, and anode and cathode gas compositions. The gas flows and the size of cell must also be equal to enable a direct comparison of the corresponding values. If not, detailed performance modeling becomes necessary to obtain performance values independent of these quantities. An approach applying impedance modeling for this purpose will be introduced in chapter 5.

4.1.2.2. LIB

For LIBs a constant current excitation changes the SOC, as pointed out in section 2.3.2.2. Instead of C/V characteristics, the performance can be analyzed by measuring the charge and discharge characteristics. A battery is subjected to charge/discharge cycles and the difference between the charge and the discharge curve gives a good measure of the performance and efficiency of the cell.

A more detailed analysis of the performance of LIBs can be found in [129]. The evaluation of the sophisticated measurement techniques like Cyclic Voltammetry (CV) and Incremental Capacity Analysis (ICA) will not be discussed in this thesis. The interested reader is referred to the references given in section 3.4.

4.2. Visualization of Impedance Spectra

Once an impedance spectrum of an electrochemical system is obtained, it has to be analyzed. Three common ways to display measured or modeled impedance spectra exist.

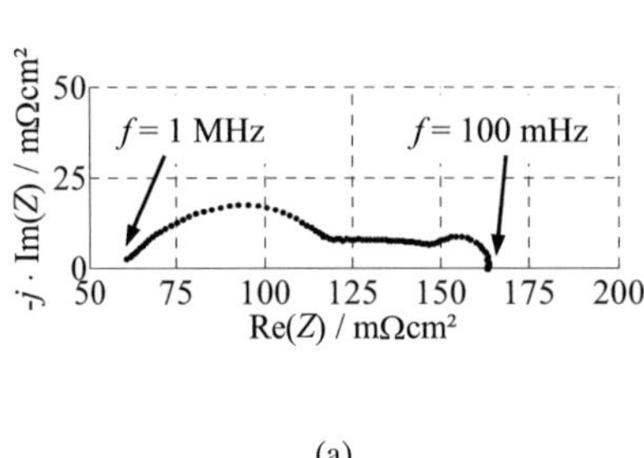

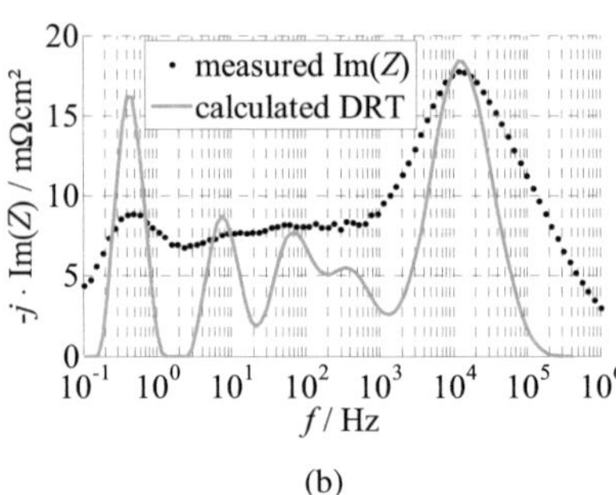

(a)   (b)

Figure 4.1.: (a) Nyquist plot of a measured impedance of a SOFC at 800 °C with a simulated reformat as fuel gas on the anode side and air as cathode gas and an
offset current density of $0.5\,A/cm^2$ (sample S.6). (b) Imaginary part of the
measured impedance shown in figure 4.1(a) (in blue) and corresponding DRT
(in red).

The one preferred by most researchers is the so-called Nyquist plot. The negative imaginary part of the impedance, $-Z''$, is plotted over the real part of the impedance, Z'. The
advantage of this display type is that an ideal polarization process is represented by a
semi-circle, often referred to as 'arc', that is easy to detect in the Nyquist plot. However,
no information about the frequency is given by a Nyquist plot itself. As an example the
Nyquist plot, one of the impedance spectra analyzed in this thesis is shown in figure 4.1(a).

The second possibility to display an impedance spectrum is the Bode plot. It consists of
two separate diagrams, the first showing the logarithmically scaled absolute value of the
impedance over the logarithmically scaled frequency, and the second showing the phase
angle either in rad or degree over the logarithmically scaled frequency.

The third possibility shown here is also presented by two individual diagrams with the
same x-axes as before, but in the first diagram showing the real part of the impedance and

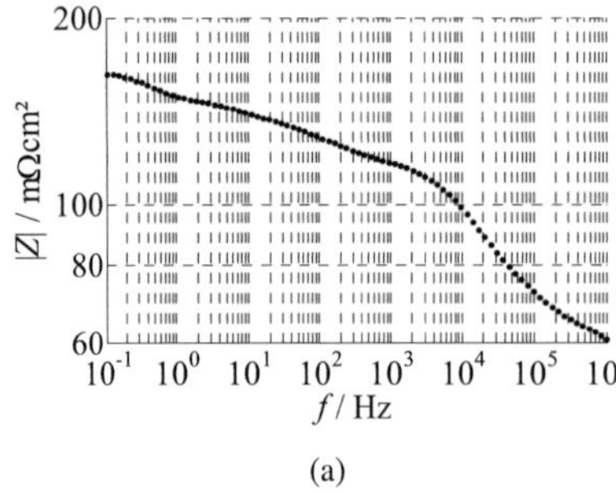

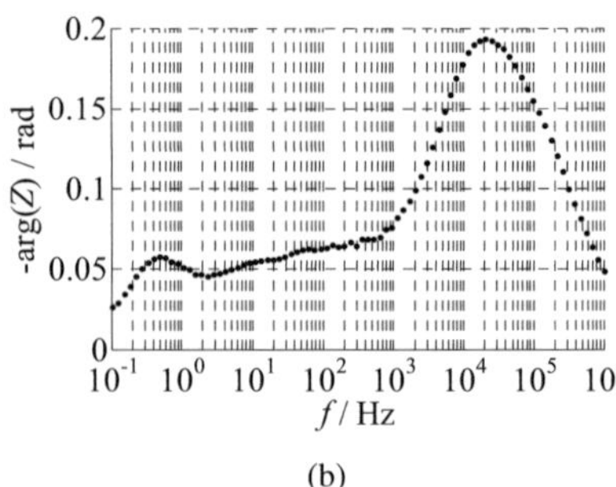

(a)   (b)

Figure 4.2.: (a) Absolute value and (b) phase angle of the measured impedance shown in
figure 4.1(a).

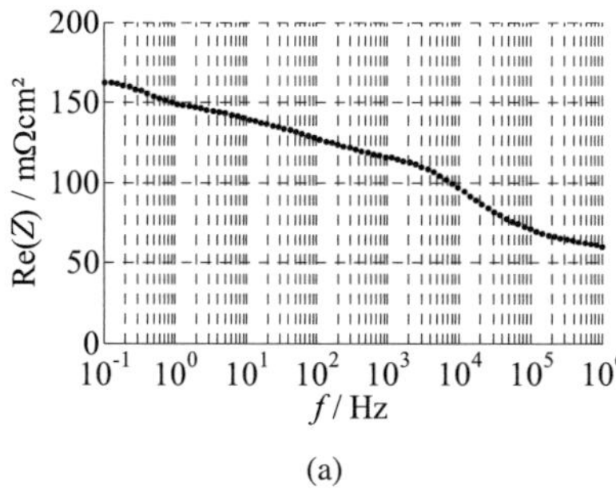
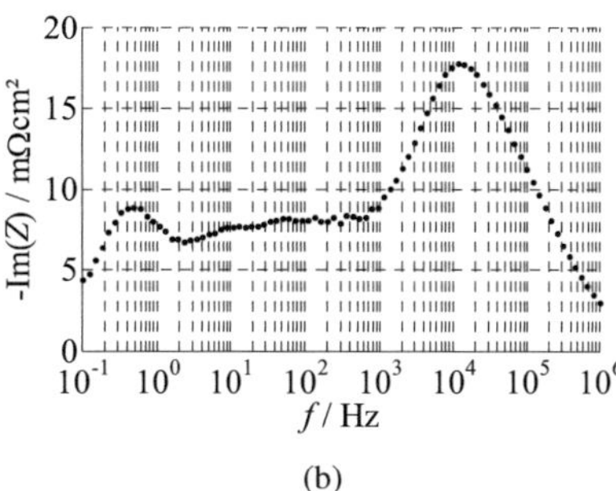

(a)   (b)

Figure 4.3.: (a) Real part and (b) negative imaginary part of the measured impedance shown in figure 4.1(a).

in the second showing the negative imaginary part of the impedance.

When comparing figures 4.2 and 4.3, it is apparent that the display types are similar. Furthermore, the absolute value and the real part of the impedance are not very helpful to identify polarization processes, whereas the phase angle and the imaginary part of the impedance clearly show two polarization processes for this example. In contrast to the Nyquist plot in figure 4.1(a), they also provide information about the characteristic frequencies of these processes. However, in the frequency range between 1 and 1000 Hz, little information about possible polarization processes can be extracted from these diagrams.

4.3. Validation of Impedance Spectra

An impedance spectrum is obtained from the Frequency Response Analyzer (FRA) as a list of complex values, one data point for every frequency that was measured. These points constitute a non-parametric model of the measured system. Nevertheless, there is a certain relation between these points, the Kramers-Kronig relation, that can be used to validate the measured impedance data.

First, general errors and deviations that are apparent without detailed analysis will be discussed in section 4.3.1. In section 4.3.2.1 the Kramers-Kronig relation will be introduced briefly. A viable technique to actually verify impedance data was proposed in [157] and will be summarized in section 4.3.2.2. Another possibility for this verification based on the DRT (see section 4.4) was proposed in [158] and will be addressed in section 4.4.1.

4.3.1. Obvious Deviations in Impedance Spectra

Displaying impedance data as Nyquist plot provides the possibility of a visual inspection of the measured data. The following aspects should be considered for every impedance spectrum.

- **Measurement noise.** A rough course of the Nyquist plot is an indicator for measurement noise, that can have various origins. The most common ones are an inappropriate measurement range, an unstable system and electromagnetic interference from other devices in the test bench or the laboratory.

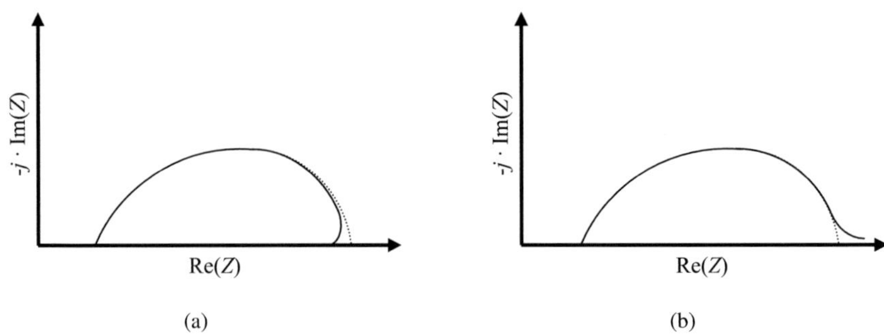

Figure 4.4.: Schematic Nyquist plots of impedance spectra showing (a) activation and (b) degradation during impedance measurement.

- **Inductive artifacts.** At very high frequencies, nearly every measurement setup produces inductive artifacts. They are indicated by a change from capacitive behavior, which is the standard behavior for a passive electrochemical system (see also section 2.3.3.2 and 2.3.3.3), to inductive behavior. If the intersection with the real axis occurs at relatively low frequencies, the measurement setup should be optimized. It should be noted that inductive effects are more pronounced for systems with low impedance values, because a larger current is needed to provoke the same voltage excitation.

- **Time variance during measurement.** There are various reasons for time variant or non-stable behavior of the tested sample. The most often encountered effects include the temperature not being in equilibrium with the surroundings, heating due to too high of a perturbation signal, and degradation or activation over time. Such a behavior is commonly detectable in the low frequent part of the spectrum, because a longer measurement time is needed there for every measurement point and time variance is more pronounced. Possible courses of the impedance are shown in figure 4.4.

Besides these, smooth and accurate looking spectra can also contain other errors. This can be analyzed by the method introduced in sections 4.3.2.1 and 4.3.2.2.

4.3.2. Kramers-Kronig Test

The Kramers-Kronig test is a powerful tool to test the quality of a measured spectrum. The theory behind this test will be explained in section 4.3.2.1. Then a practical realization of this test for finite measurement data is presented in section 4.3.2.2. Another possible method to obtain a quality criterion related to the Kramers-Kronig test with the help of the DRT is shown in section 4.4.1.

4.3.2.1. Kramers-Kronig Relation

Consider a measured impedance represents a dynamic system, that fulfils the following criteria:

- **Causality.** The measured response signal is exclusively caused by the perturbation.

- **Linearity.** The system was measured in the linear regime.

- **Stability.** The sample did not change its electrochemical properties during the measurement.

- **Finiteness.** The impedance values must be finite for both $\omega \to 0$ and $\omega \to \infty$. There are two exceptions for this criterion:

  - Purely inductive behavior for $\omega \to \infty$ is not finite yet constitutes a valid impedance.

  - A LIB shows capacitive behavior $(\mathrm{Im}(Z(\omega \to 0))) \to -\infty)$ and is also part of a valid impedance spectrum.

In this case the impedance complies to the Kramers-Kronig relation. This means, the imaginary part of the impedance can be reproduced by the real part of the impedance and vice versa. This has been derived and explained in many publications [36, 122], where the equations of the Kramers-Kronig relation are also presented.

A practical way of illustrating the Kramers-Kronig relation is a simple RC circuit (see section 2.3.3.2). The course of the real part of the impedance is a function of the parameters R_{RC} and C_{RC}. Since this also accounts for the imaginary part, the two courses are connected by a mathematical transformation. This is similarly the case for every impedance response that fulfils the above mentioned criteria.

4.3.2.2. Kramers-Kronig Test by Boukamp

The Kramers-Kronig relation has been explained for an RC circuit in section 4.3.2.1. As the impedance response of an RC circuit complies the criteria for the Kramers-Kronig relations, so too does every impedance response that can be reconstructed by a series connection of RC circuits. This is the idea behind the Kramers-Kronig test proposed in [157].

In order to conduct this test, a number of RC circuits, n, has to be chosen. The values of the characteristic time constant, τ_n, must also be predefined. They are usually logarithmically spaced and cover the frequency range of the measurement. Then the ECM (see section 4.5) consisting of the series connection of the n RC circuits plus a resistor R_0 is fitted to the measurement data. Because the value for each τ_n is fixed, the fit is linear in its parameters R_n and R_0.

For systems with capacitive behavior (like LIB) or if pronounced inductivities are visible in the tested spectrum, the ECM can be expanded by a capacitor or an inductor, respectively. These elements violate the criterion of finiteness, but can be caused by a causal, linear, and stable system, as stated in section 4.3.2.1.

The residuals between measured and fitted ECM give a measure of the applicability of the Kramers-Kronig relations and are called Kramers-Kronig residuals.

- **Low residuals** indicate good measurement quality and that the criteria listed in section 4.3.2.1 are complied.

- **Randomly distributed residuals** show that a certain measurement noise is present during measurement, but the criteria in section 4.3.2.1 are basically satisfied.

- **Systematically distributed residuals** can occur for two reasons: a serious violation of the criteria for the Kramers-Kronig relations or the number of RC circuits n was chosen too small (see below).

Simulations with different numbers of RC circuits showed that:

- a low number for $n < n_{\mathrm{opt}}$ leads to poor fit results with large systematic deviations, i.e. large systematic residuals,

- at a certain number n_{opt} the residuals converge to a certain level,

- for a larger numbers of RC circuits $n > n_{\mathrm{opt}}$ the values for R_n start oscillating around zero. This is also stated in [157]. It is further argued, that this does not violate the test criterion.

It should be noted that the number of RC circuits is the crucial parameter for this test. Residuals for different measurements are only to be compared, if the fit was conducted with the same number of elements. It is further recommended to determine n_{opt} iteratively. This test has been implemented in *Matlab* at IWE.

4.4. Distribution of Relaxation Times (DRT)

A Debye polarization process is fully characterized by its time constant τ and its polarization loss R [21]. In an ECM, this process is represented by an RC circuit, as introduced in section 2.3.3.2. Certain processes with relaxation times that are distributed around a main value can be described by a parallel combination of an ohmic resistance and a Q element (see section 2.3.3.5). When choosing an ECM, one assumes a certain number and type of dispersion processes in the cell, which is often subject to ambiguity. Several ECMs must be tried until an adequate ECM is found. An alternative to this indirect approach is to use an arbitrary DRT. This corresponds to a 'general' equivalent circuit consisting of an infinite number of 'differential' RC circuits in series. The measured impedance $Z(\omega)$ can then be expressed by an integral equation containing the distribution function $\gamma(\tau)^2$:

$$Z(\omega) = R_0 + R_{\mathrm{pol}} \int_0^\infty \frac{\gamma(\tau)}{1 + j\omega\tau}\mathrm{d}\tau, \tag{4.1}$$

where R_0 represents the ohmic resistance, and R_{pol} is the total polarization resistance as shown in figure 2.15(a). The expression $\gamma(\tau)/(1 + j\omega\tau)\mathrm{d}\tau$ specifies the fraction of the overall polarization with relaxation times between τ and $\tau + \mathrm{d}\tau$. This implies that the area under each peak in the DRT equals the polarization resistance of the corresponding loss mechanism. In practice, the continuous function $\gamma(\tau)$ is approximated by the discrete function

$$Z(\omega) = R_0 + R_{\mathrm{pol}} \sum_{n=1}^{N} \frac{\gamma_n}{1 + j\omega\tau_n} \tag{4.2}$$

---

^{2}In the literature, it is often distinguished between the two distribution functions $\gamma(\tau)$ and $g(\tau)$ [36, 159], the former being the analytically calculated and the latter being the numerically approximated one. However, in this thesis, only the approximated distribution will be used with the denomination $\gamma(\tau)$.

with predefined values for τ_n, which are logarithmically distributed as a standard. For reasons of compatibility with the DRT software package *FTIKREG* [160],

$$\sum_{n=1}^{N} \gamma_n = \frac{1}{\Delta_{\ln}\tau_n} \quad \text{with} \quad \Delta_{\ln}\tau_n = \ln \tau_{n+1} - \ln \tau_n, \quad n = 1, \ldots, N-1 \tag{4.3}$$

is used. The calculation of the values for γ_n is not a trivial task [36]. It can performed via Tichonov regularization [161, 162] and yields good results for the approximation of $Z(\omega)$ in equation (4.2). More detailed information about the DRT can be found in [18, 36].

The benefit of this function can be observed in figure 4.1(b), where the negative imaginary part of the impedance taken from figure 4.1(a) is compared to the DRT calculated from the identical impedance. In this figure the DRT exhibits the two main polarization processes that can also be identified by the imaginary part of the impedance (cf. figure 4.3(b)) at approximately 0.4 and 10000 Hz. However, in the frequency range between 1 and 1000 Hz, where no distinct information about the involved polarization processes can be extracted from the imaginary part of the impedance, the DRT clearly shows three separated peaks corresponding to three polarization processes. This characteristic is beneficial for the analysis of impedance spectra, which will be demonstrated in section 4.5.4.

4.4.1. Kramers-Kronig Test with the Help of the DRT

When comparing the γ_n values for the DRT in section 4.4 and the R_n values for the Kramers-Kronig test in section 4.3.2.2, it becomes apparent that the only difference is the method in which these values are obtained, the Tichonov regularization or a linear fit, respectively. Therefore, the DRT method yields a smooth curve, whereas the values for R_n plotted over τ do not.

For any DRT, a back calculation with the help of equation (4.2) can be performed. The result is $Z_{\mathrm{mod}}(\omega)$. A residual plot comparing measured impedance $Z(\omega)$ with the impedance calculated from the corresponding DRT $Z_{\mathrm{mod}}(\omega)$ yields similar information as the Kramers-Kronig test and can also be consulted to rate the quality of the spectrum with the same criteria as the Kramers-Kronig test in section 4.3.2.2. This is discussed in detail in [157].

4.5. Equivalent Circuit Modeling

An ECM tries to reproduce the impedance response of an electrochemical system with a composition of electrical equivalents (see also sections 2.3.3.2 and 2.3.3.5). The ECM should always contain the minimum number of circuit elements possible to represent the measurement data adequately – this facilitates both the parameter identification and the physical interpretation of the ECM. At best, each subcircuit should represent an electrochemical process. An accurate parameterization yields important information about this process, as demonstrated in figure 4.5:

- **The polarization resistance of a subcircuit** represents the portion of the whole polarization resistance, R_{pol}, that this particular loss process is responsible for.

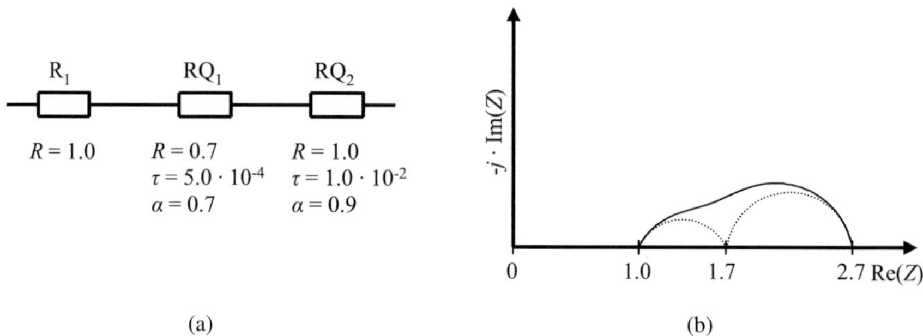

Figure 4.5.: (a) Example of an ECM consisting of a resistor R and two RQ circuits. (b) Impedance of the ECM from figure 4.5(a) depicted together with the impedance of the single impedance contributions.

- **The characteristic frequency** yields valuable information about the speed of the corresponding process and can thereby help to identify a physical effect behind it.

- **The attainable precision** gives a measure of how adequate the ECM actually is.

The stepwise identification of the loss processes of a SOFC single cell by means of a physically motivated ECM is demonstrated in [18]. The evaluation of a whole measurement series over time can attribute the degradation to a specific loss process [91].

The parameterization of an ECM is commonly done by fitting the parameters of a pre-designed ECM to impedance data obtained from measurement, as will be explained in section 4.5.1. Since this is not a trivial task, sections 4.5.2 to 4.5.4 present three possibilities to facilitate the fit procedure. Different fit algorithms for the different approaches are discussed in section 4.5.6. In section 4.5.5 a batch fitting approach is presented which is based on the methods derived in sections 4.5.2 to 4.5.4. With this approach, it is possible to parameterize a complete measurement series automatically.

4.5.1. Complex Nonlinear Least-Squares Fit

In the case of a linear fit problem, the quality criterion is a quadratic function with one global minimum that can be calculated analytically [137]. But even for the fit of the most simple subcircuit representing a polarization process – the RC circuit – the fit problem is nonlinear and complex. It is nonlinear because the impedance $Z(\omega)$ exhibits a nonlinear dependency on both parameters R_{RC} and C_{RC}. It is further complex because the quality criterion consists of two parts – one for the real part of the impedance and one for the imaginary part of the impedance. Generally, the quality criterion is given by:

$$S = \sum_{m=1}^{M} \left[w_{\mathrm{Re},m}(\mathrm{Re}(Z(\omega_m)) - \mathrm{Re}(Z_{\mathrm{mod}}(\omega_m, \mathbf{a})))^2 \right.$$

$$\left. + w_{\mathrm{Im},m}(\mathrm{Im}(Z(\omega_m)) - \mathrm{Im}(Z_{\mathrm{mod}}(\omega_m, \mathbf{a})))^2 \right], \quad (4.4)$$

where the vector a contains all free parameters of the fit problem. The main challenge of the CNLS fit is the consistent minimization of the quality criterion defined by equation (4.4). The initial values for the fit play a very important role in the fitting procedure.

These have to be more accurate the more complicated the ECM is. Only in this way can it be guaranteed that the fit converges in the global minimum and not in a local one. Special algorithms are also needed for this task: two will be introduced in section 4.5.6. A detailed introduction into CNLS fit and guidelines for how to choose the fitting parameters (number of iterations, weighing functions, complex, real only or imaginary only fit) is given in [163]. A recommended and widely used software tool is *Zview*. Its manual also gives a good overview of CNLS fitting and the independent quality criteria described here. Nevertheless, for the fits in this thesis, a powerful new technique is presented in section 4.5.4.

4.5.2. Pre-Identification via DRT

As demonstrated in figure 4.1(b), the DRT is a very good type of visualization for displaying the individual processes occurring in the electrochemical system. This is why a pre-identification via DRT is a valuable tool for evaluation of the number of processes measured in the impedance spectrum. As a rule, it can be stated that if two or more expected processes do not appear separated in the DRT, they are unlikely to be deconvolutable via CNLS fitting. In some cases, processes can be visualized by the DRT that are not visible in the Nyquist plot, as they are hidden by larger processes or because a number of processes appear as one smeared process in the Nyquist plot (cf. figure 4.1(b)). The DRT can thus be consulted in order to develop an appropriate ECM, as was done in [18].

Once an ECM is determined, the DRT can be used to identify suitable initial values for the CNLS fit. First, the characteristic frequencies of the polarization processes can be identified via the DRT. This can also be done using the imaginary part of the impedance, since the peaks indicate the characteristic frequencies. However, a higher degree of separation of the peaks in the DRT, as shown in figure 4.1(b), enables easier identification of the peak frequencies for complicated systems. As mentioned before, the area under each peak is proportional to its fraction of the total polarization resistance. With this information, initial values for the resistances in the ECM can also be estimated.

4.5.3. Consistency Check of Fit via DRT

The DRT also yields the possibility to double-check the obtained values from the fit procedure. When calculating the DRT of the impedance of the ECM parameterized with the obtained fit result and comparing this DRT with the DRT of the measurement, systematic errors like invalid characteristic frequencies or deviating resistances can be visualized as demonstrated in figure 4.6. The measured impedance from figure 4.1 is shown again together with a corresponding fit result that was obtained with rather bad initial values by *Zview*. The fit result reproduces the measured impedance in an acceptable way, but the DRTs of these impedance curves in figure 4.6(b) dramatically reveal that the obtained parameter set does not represent a valid solution of the fit problem. Therefore, it is proposed to double-check the obtained parameter sets with this method before using them for further calculations.

4.5.4. DRT as additional Quality Criterion

The surplus of information of the DRT can also be used directly in the fit algorithm itself. It has to be kept in mind that all information for the calculation of the fit result of the

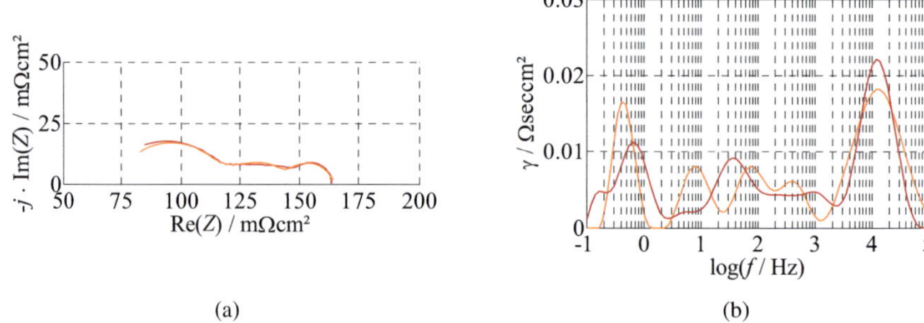

(a)　　(b)

Figure 4.6.: (a) Nyquist plot of measured impedance from figure 4.1 (in red) and corresponding fit result obtained by *Zview* with bad initial values (in orange). (b) DRTs of measured impedance from figure 4.1 (in red) and of fit result shown in figure 4.6(a) (in orange).

standard fitting procedure is the real and the imaginary part of the impedance as shown in figures 4.1 to 4.3. This may work out well for clearly separated processes like the two mentioned in the description of figures 4.1 to 4.3 in section 4.2, but it is apparent that a clear assignment of characteristic time constants and corresponding resistances for the frequency range between 1 and 1000 Hz is not trivial. The data of the DRT depicted in figure 4.1(b) seems to be more appropriate to perform this task. As mentioned in section 4.5.3, a comparison of the DRTs of measured and modeled impedance can yield additional information about the quality of the fit. The idea behind the technique proposed in this study is to expand the quality criterion for the fit by the difference of these two curves. This results in the new quality criterion:

$$S = \sum_{m=1}^{M} \left[w_{\mathrm{Re},m}(\mathrm{Re}(Z(\omega_m)) - \mathrm{Re}(Z_{\mathrm{mod}}(\omega_m, \mathbf{a})))^2 \right.$$

$$\left. + w_{\mathrm{Im},m}(\mathrm{Im}(Z(\omega_m)) - \mathrm{Im}(Z_{\mathrm{mod}}(\omega_m, \mathbf{a})))^2 \right]$$

$$+ \sum_{n=1}^{N} \left[+ w_{\mathrm{DRT},n}(\gamma_n - \gamma_{n,\mathrm{mod}}(\mathbf{a}))^2 \right]. \quad (4.5)$$

Because of the increased complexity of the quality criterion, the Levenberg-Marquardt algorithm (see section 4.5.6.1) fails for this task and was substituted by a simplex algorithm (see section 4.5.6.2). Figure 4.7(a) shows the obtained fit result with the new quality criterion and the new fit algorithm. For the fitting procedure, the same initial values were taken as for the fit shown in figure 4.6(a). It can be observed that not only is the Nyquist plot in figure 4.7(a) reproduced with good accuracy, but also the comparison of the corresponding DRTs shows little deviation. As can be seen from the residual plot in figure 4.8(a), the only significant deviation occurs for very high frequencies and can be interpreted as measurement artifact due to inductances that originate from the cell connections at high frequencies. The inductive behavior is not modeled in the applied ECM and therefore appears as a deviation in the residual plot. Another minor deviation can be observed at

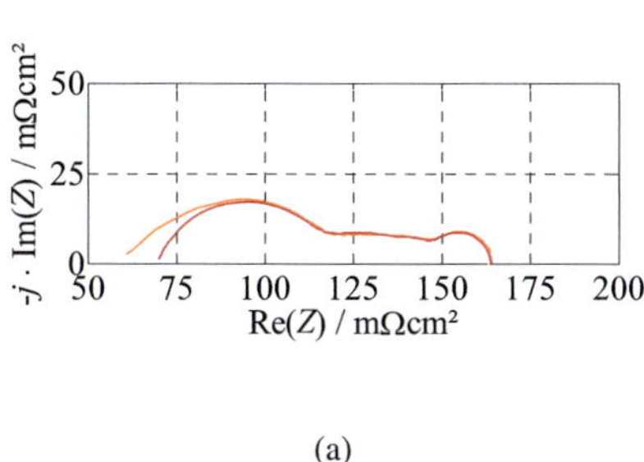 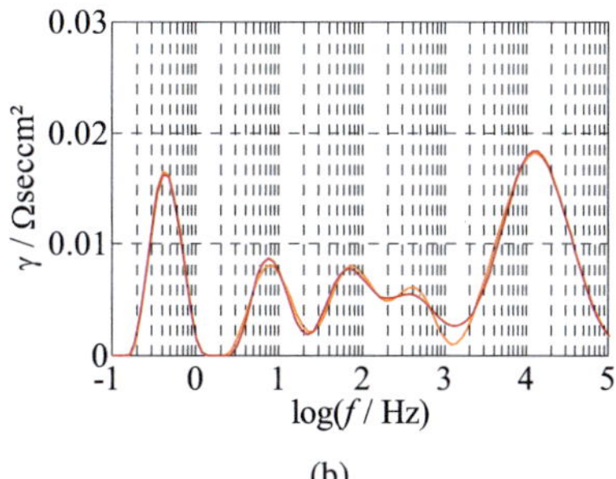

(a) (b)

Figure 4.7.: (a) Nyquist plot of measured impedance from figure 4.1 (in red) and corresponding fit result obtained by by the here proposed fitting procedure with bad initial values (in orange). (b) DRTs of measured impedance from figure 4.1 (in red) and of fit result shown in figure 4.7(a) (in orange).

around 1000 Hz. This is the frequency above which a special high pass filter in the measurement device is activated. The noise between 70 and 1000 Hz is due to a characteristic noise in the imaginary part of the impedance that is observed frequently for such low absolute impedance values in this frequency range. Again, it can be stated that the deviation results from a measurement artifact and not from a poor fit result. This can also be shown by the similar residuals in figure 4.8(b) obtained by the comparison of the measured impedance in figure 4.1(a) and the impedance calculated from the corresponding DRT in figure 4.1(b), as explained above.

Figure 4.8(b) shows, that the deviation of the measurement from the causal, stable, and linear system determined by the DRT[3] are of the same quality and appear in the same frequency range as the fit residuals in figure 4.8(a). This means that the fitting procedure has even compensated for measurement artifacts revealed in figure 4.8(b). This robust-

---

[3]The ability to validate impedance data via the DRT has been subject to section 4.4.1.

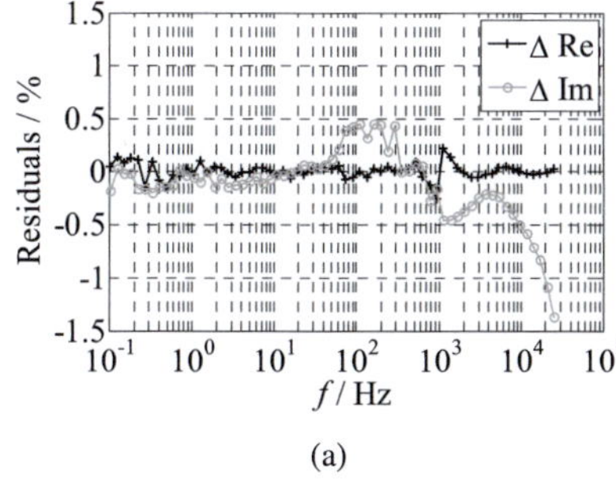 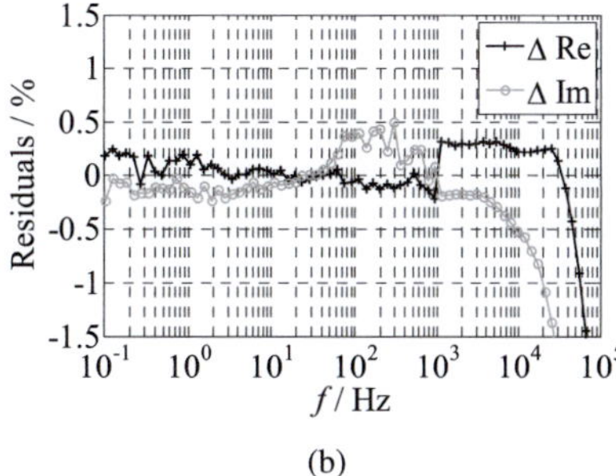

(a) (b)

Figure 4.8.: (a) Residuals of the fit result shown in figure 4.7(a). (b) Residuals of the measured impedance (figure 4.1(a)) and the impedance calculated from the corresponding DRT (figure 4.7(b)).

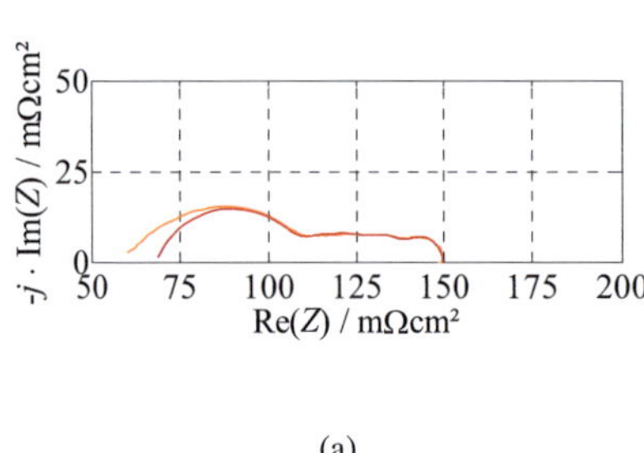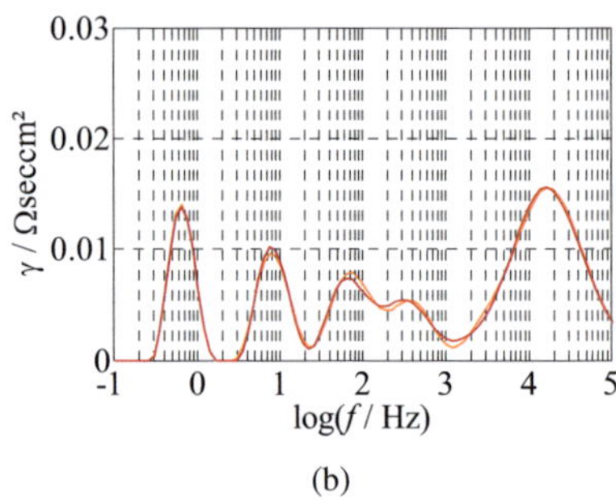

(a)   (b)

Figure 4.9.: (a) Nyquist plot of the same cell as in figure 4.1(a) but with an offset current density of $J = 1.0\,\text{A/cm}^2$ (in red) and corresponding fit result (in orange) obtained by the here proposed fitting procedure with the fit result shown in figure 4.7(a) as initial guess. (b) DRTs of measured impedance (in red) and of fit result (in orange) shown in figure 4.9(a).

ness is highly beneficial for a consistent batch-fitting for measurement series, as will be explained in section 4.5.5.

4.5.5. Batch Fitting

A batch-fitting procedure yields fit results not only for one spectrum but a whole measurement series, normally a variation of one operating parameter or simply the measurement time. After fitting the first spectrum of this series, the fit result is taken as an initial guess for fitting the next spectrum in line and so forth. It is obvious that a very robust fitting procedure is needed in order to guarantee consistent fitting throughout the whole series, as a deviation in one of the first spectra is not likely to be compensated in later fits. In figure 4.9(a) the Nyquist plot of a spectrum with increased offset current density of $J = 1.0\,\text{A/cm}^2$ (compared to $J = 0.5\,\text{A/cm}^2$ in figure 4.1(a)) is depicted together with the corresponding fit obtained by the proposed method. The fit result shown figure 4.7(a) was taken as initial value.

Both impedance and the DRTs show little deviations and enable the fitting of another spectrum measured at a further increased current density of $J = 1.5\,\text{A/cm}^2$ with the same procedure. The results are shown in figure 4.10.

Again these figures show a very good accordance in fit and DRT, apart from the inductive effects that cannot be reproduced because of the lack of a corresponding model. In figure 4.11(a) the whole measurement series together with the fits plus the corresponding residuals are depicted. Apart from the deviation due to the inductive artifacts, the residuals are very small with similar characteristics as the residuals shown in figure 4.8(a). In fact, the characteristic deviations that were discussed in the previous section appear again in this residual plot. This underlines the statement that these deviations can be explained by measurement artifacts.

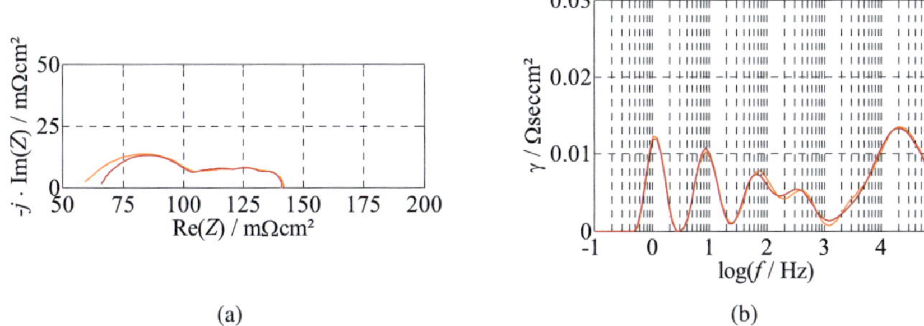

(a)   (b)

Figure 4.10.: (a) Nyquist plot of the same cell as in figures 4.1(a) and 4.9(a) but with an
offset current density of $J = 1.5\,\text{A/cm}^2$ (in red) and corresponding fit results
(in orange) obtained by the here proposed fitting procedure with the fit result
shown in figure 4.9(a) as initial guess. (b) DRTs of measured impedance (in
red) and of fit result (in orange) shown in figure 4.10(a).

4.5.6. Fit Algorithms

In this section, two optimization algorithms are introduced briefly. The Levenberg-
Marquardt algorithm in section 4.5.6 is a widely used algorithm for CNLS fitting problems. In section 4.5.6.2 the Nelder-Mead method is presented. This method is often also
called downhill simplex method or simplex algorithm and is a search algorithm without the need of derivatives. It is therefore well applicable for the method described in
section 4.5.4.

4.5.6.1. Levenberg-Marquardt

The Levenberg-Marquardt algorithm is often applied for CNLS fitting problems [22].
This algorithm yields good fit results, as it combines the advantages of steepest gradient

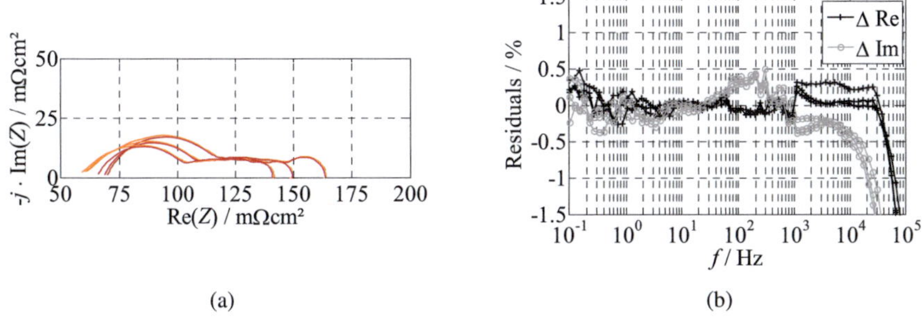

(a)   (b)

Figure 4.11.: (a) Measurement series as shown in figures 4.1(a), 4.9(a) and 4.10(a) with an
offset current density of $J = 0.5, 1.0, 1.5\,\text{A/cm}^2$ (in red) and corresponding
fit result (in orange) obtained by the here proposed fitting procedure. (b)
Residuals for the three fits shown in figure 4.11(a).

descent when far away from the minimum, and of the Gauss-Newton-algorithm for good and fast convergence when near the global minimum [164, 165, 166]. The search direction is actually a cross between the Gauss-Newton direction and the steepest descent direction. This algorithm yields very accurate and fast convergence if the Jacobian matrix for all fit parameters is provided. If not, the algorithm can calculate them numerically but thereby loses performance.

4.5.6.2. Nelder-Mead Method or Simplex Algorithm

A simplex is the convex hull of $n + 1$ verteces in the n-dimensional space. It is the minimum representation of a full-dimensional object, i. e. a triangle in the plane or a tetrahedron in the 3-dimensional space.

The Nelder-Mead method starts with a simplex in the parameter space. By searching external positions with lower values for the quality criterion and by canceling the vertex with the highest value for the quality criterion, the simplex moves towards the minimum of the quality criterion. If no favorable external position can be found, the simplex is refined and shrinks.

This way it is possible to find the solution of a nonlinear optimization problem without calculating the Jacobian matrix [167, 168].

4.6. Other Analysis Techniques for Impedance Spectra

A number of other possibilities exist to extract more information out of impedance spectra. Differential Impedance Analysis (DIA) calculates a new function out of the real and imaginary parts of the impedance plus their first and second derivatives [169]. However, the calculation of derivatives of measurement data is always suspect, as noisy data leads to large artifacts in the derivative. This is probably the main reason why this method is not widely used in electrochemistry.

In [170] it is proposed to plot the difference in the imaginary part of two impedance spectra, provided that a measurement series of impedance spectra has been recorded upon variation of one specific operating parameter. At best this variation should trigger only one polarization process. The polarization processes can then be analyzed in detail by the proposed technique. This method has been called Analyzing the Difference in Impedance Spectra (ADIS) in [171] and was expanded to a more sophisticated technique by introducing some rules for its application. One of the spectra, preferably that for the minimum or the maximum of the varied operating parameter, is declared as reference and the imaginary part of this spectrum is subtracted from all other spectra. This yields quite a good visualization of the consequences of the parameter variation and can help to identify different polarization processes much better than with a single spectrum. However, the corresponding polarization process must be sensitive enough to produce a detectable change. This may not be possible for some processes. The effort for this method is also quite high, as a whole measurement series is needed and not only a single impedance spectrum.

5. Impedance Modeling of Electrochemical Systems

In the preceding chapters 3 and 4 methods have been introduced to obtain important properties of electrochemical systems. The methods are suitable for small-scale systems and can yield important information about the performance and efficiency of the tested system as well as their dependencies on operating parameters.

However, it is important to note that a small-scale system is a model system designed for characterization purposes in a laboratory environment and not intended for commercial application. Large-scale systems are needed to provide enough power or energy to be technically relevant. The differences between small- and large-scale systems will be elucidated in section 5.1. Accordingly, different approaches and model types are needed for both types of systems, namely micro-models and macro-models.

The research group of Bessler focuses on the so-called 'multi-scale model' approach, where even effects on the atomistic scale are considered [105, 172]. The parameters are then inserted into the next scale and homogenized for faster calculation. This scale-up step is repeated until the desired system level is reached. Among others, impedance models are also proposed [96].

In this thesis, a more self-contained approach based on impedance models will be presented. Impedance models are a good basis for a modeling framework, as they combine static and dynamic behavior of the system with a short calculation time suitable for real-time simulation in the time domain. In section 5.2 impedance models will be introduced. An approach to apply these impedance models to macro-models is subject to section 5.3.

Hence, the main content of this chapter is the introduction of an impedance model framework that is applicable to simulate a Solid Oxide Fuel Cell (SOFC) stack statically and dynamically. The static model is appropriate to give an estimate for the expected power output of a SOFC stack. It does not require measurements on complete stacks. It is sufficient to parameterize the model only with single cell measurements on A_1 cells (see figure 3.1). These measurements must include the complete relevant operation parameter field which will be present at any position in the stack. It will be demonstrated by a comparison of modeling and measurement that the model parameterized with A_1 measurements produces accurate results for the simulation of an A_{16} cell for different operating points.

As will be shown in section 5.3.2, the individual losses of a SOFC can only be separated with the help of dynamic measurements like Electrochemical Impedance Spectroscopy (EIS) and corresponding analysis methods. In a micro-model a change in the impedance spectrum can easily be attributed to an individual loss process, if an adequate Equivalent Circuit Model (ECM) is available [91]. For stacks this task is more difficult due to the difficult technical realization of dynamic measurements on the stack and the lack of appropriate dynamic models for SOFC stacks.

However, it will be shown that the dynamic micro-model can also be scaled up to identify the electrochemical losses (section 5.3.5)[1] and – within limits that will be discussed – reproduce the impedance of a larger cell (section 5.3.6). Attention must be turned to the Gas Conversion Impedance (GCI) which only needs to be considered for large-scale systems. This will be discussed in detail in section 5.3.7.

The model framework presented in this chapter is aimed for application in a diagnosis system for SOFC stacks, but it also allows to predict the stack performance out of single cell measurements with minimal computational effort.

It should further be mentioned, that this approach is primarily intended for SOFCs in H_2/H_2O operation. The framework itself is not restricted to this setup, but the model presentation in section 5.3 will leave out Lithium-Ion Batterys (LIBs) and other fuel gas compositions for SOFCs in order to keep the presentation comprehensive. For the obtained model the SOFC and its particularities as modeled electrochemical system play also an important role.

5.1. Small/Large-Scale Systems and Micro/Macro-Models

Simple model systems are generally used for the identification of the governing processes taking place in electrochemical systems, as the number of involved processes should be as small as possible for their identification and subsequent modeling. On the other hand, it is sensible to characterize technically relevant structures, because the applied materials may show altered properties when embedded in a complex system compared to their behavior in model systems [66].

This chapter concentrates on SOFCs as a practical example for these proposed methods. As pointed out above, A_1 cells are used to model small-scale system behavior, on which the identification of the relevant processes is conducted. A short definition of small-scale systems and the corresponding model is given below.

- **Small-scale system.** A small-scale system is the smallest sensible device that allows for a complete characterization of all individual processes in a given electrochemical system. The only difference to technically relevant systems is its size. One benefit of reduced size is that homogeneous operating conditions can be easily

---

[1] This section is based on the results already published in [173].

adjusted over the whole electrochemically active area. Other advantages are reduced costs compared to larger systems and the easier fabrication of samples with homogeneous properties.

- **Micro-model.** The corresponding model to a small-scale system taking into account all individual processes of a given electrochemical system. Only spatially static operating parameters are used as input parameters and no spacial distribution of output quantities is provided.

The range of possible model approaches is quite large for small-scale systems. But for technically relevant systems with distributed parameters, these are often not applicable due to the increased complexity. This comprises spatially distributed input and output parameters and the interaction of small sections within the system with the rest it. Hence, technically relevant systems are treated as another category of systems and the terminology for these large-scale systems as well as the corresponding macro-models is introduced in the following.

- **Large-scale system.** A large-scale system is an electrochemical system, in which all possible processes and reactions take place, that are supposed to take place in a commercial electrochemical system of the same type such as a SOFC stack. All relevant quantities have to be considered as distributions in all relevant dimensions. The size and complexity of these systems reach from large laboratory style test cells to full commercial products or prototypes.

- **Macro-model.** This is a model for a large-scale electrochemical system and takes into account all processes likely to occur in a commercial product. All relevant quantities are given as distributions in all relevant dimensions. It can be constructed on the basis of a small-scale model but has to be expanded by one or more dimensions. Usually, Finite Elements Method (FEM)-based models are designed for this purpose.

5.1.1. Small-Scale Systems

The purpose of these systems has already been mentioned. The micro-models developed for these model cells can be stunningly precise and help to understand their behavior. However, the general rule is that the more precise the model, the more complex it usually is. Because of the large differences between small-scale systems for SOFCs and LIBs, they are discussed separately in the following subsections.

5.1.1.1. SOFC

In SOFC research, single-cells of different dimensions are common. But further simplified test samples are manufactured to refine identification of fundamental processes in these cell configurations. For example model anodes can be used to identify the processes on the anode on the atomic level [75]. Symmetrical cathodes are selected to measure and analyze the electrochemical reaction on the cathode side of the SOFC [10, 31, 135]. Measurements can be conducted in a much simpler test setup, because ambient air can be supplied to both sides of the sample and therefore traditional sources of errors such as gas tightness are not an issue. However, these model systems are not full cells, and they are not considered in this thesis.

The single-cells used at Institut für Werkstoffe der Elektrotechnik (IWE) as small-scale systems have a square-shaped active electrode area of $A_1 = 1\,\mathrm{cm}^2$ (see figure 3.1). This is seen as good trade-off between a tolerable variation of operating conditions along the gas channel and negligible edge effects. But a small single-cell does not ensure homogenous operating conditions alone. The gas supplied has to be uniformly provided and the flow must be adequately large. Only under these preconditions, a reproducible measurement can be achieved and guaranteed.

5.1.1.2. Li-Ion Batteries

Button cells are often used to characterize components of Li ion batteries. They have a small electrode area and the housing shows a good thermal conductivity, so that every part of the cell is well connected to the current collector and the operation temperature is kept equal all over the active area. The *EL-Cell* offers special re-usable cell housings with an electrode area of $A = 2.54\,\mathrm{cm}^2$. Also so-called *Swagelok* cells are used. Here gas pipe adaptors are used as housings, taking advantage of the their gas tightness and robustness. However, it has turns out that several problems arise with this type of housing. Corrosion, long term degradation, and lacking reproducibility were drawbacks in attempts to apply these housings for basic research. Unfortunately, such results seldom get published in the literature, so no reference can be given for these findings.

5.1.2. Large-Scale Systems

However accurate a micro-model can get, a technically relevant system will most likely not fulfil the requirements to assume constant operating parameters. For example the core temperature of a prismatic LIB is supposed to be higher than its surface temperature, due to the losses (see section 2.4.2) and the entropy change during cycling.

The system with homogeneous operating parameters becomes a system with distributed parameters, as it is expanded by at least one additional dimension. These are the dimensions, in which the change in relevant operating parameters occurs. Examples of large-scale systems relevant for this thesis can be found in the following subsections.

5.1.2.1. SOFC

The most prominent and also most often used commercial system is the so-called SOFC stack. Several planar layers are mounted on top of each other, separated by a bipolar plate that provides electrical connectivity and separates the gas volumes [11]. In a first approximation the gas supply and operating temperature can be assumed equal for all gas channels in the stack, but a non-negligible variation of operating parameters occurs along these gas channels as will be explained in section 5.3.4.

Many researchers and industrial companies focus on another type of a SOFC: tubular cells. As they are not subject to this thesis, the reader is referred to [174] for a good introduction into manufacturing, characteristics and performance of such systems.

5.1.2.2. LIB

As the capacity of a LIB scales with the cell area, commercial LIBs require a large cell area. The approaches to achieve this include the three most often used designs for LIBs: pouch cells, prismatic cells and round cells (see also section 2.2.1.4). But no adequate reference can be given for a critical discussion about small-/large-scale systems and micro-/macro-models at this point.

5.1.3. Micro-Models and Macro-Models

As stated in section 5.1.1, even small-scale models can be quite complex. The structure is commonly defined by the quality (type and complexity) of the governing equations. Therefore a scale-up of models for technically relevant systems is not an option for most modeling approaches.

However, the identified electrochemical relations apply also for large-scale systems. A classical way to implement a model that accounts for a spatial distribution of operating parameters is a FEM model. In this approach, the governing equations are evaluated at special points in a mesh that is generated in the geometrical structure of the system. FEM simulations are capable of providing a large number of physical and chemical quantities at each position inside the model and can therefore give a good insight into the system, but they require a complex software framework such as *Comsol Multiphysics*.

The macro-models presented here are constructed in a different way. They take the micro-model as basis and expand it by adding distributed variables, that account for the shift in operation parameters within the large-scale system. The most prominent example of such an expansion is the gas channel in a SOFC system. Here, one additional dimension is added to account for the changes in operating conditions alongside this dimension. This example will be discussed in detail in section 5.3.

Macro-models that can be found in the literature are commonly designed for a single characteristic or physical quantity. Often the whole cell area is homogenized and one single Area Specific Resistance (ASR) is used for both ohmic and polarization resistance [175].

5.2. Introduction into Impedance Modeling

The physical and chemical processes that occur in electrochemical systems have many different natures and comprise many research areas. However, if they are relevant during operation, they will produce a voltage response for a given current excitation and vice versa, because of the inhibited impedance.

That said, the impedance of an electrochemical system can be used for two major purposes:

- **Identification.** It was already shown with the help of the ECMs in section 4.5, that impedance analysis is a substantial means for the identification of electrochemical processes occurring in the system.

- **Simulation.** Furthermore, an impedance model[2] is usually a very simple mathematical expression that reflects a deterministic dynamic relation between voltage and current density. Within a specific operation area, it is easily applicable for a time domain simulation of the system, as required for various purposes [109]. If a linear operating area cannot be assumed, the parameters of the ECM can be adjusted according to the current operation area in time, resulting in a time variant system.

The first of these purposes has long been known and long been applied for a large variety of electrochemical and other technical, dynamic systems.

The second one becomes more and more important, as time domain simulations are a classical way to simulate the powertrains of automotive vehicles – one of the largest industrial sectors worldwide. Within the last ten years, with the development of fuel cells and batteries, large car manufacturers try to include electrochemical systems in their time domain powertrain simulations. Thus, in parallel with the starting commercialization of these systems for automotive applications, the need for fast and easily implementable models for time domain simulations of fuel cells and batteries arises.

As already stated, impedance models are quite simple in their structure and apply only a small number of parameters. They have the potential to be applied for big systems like battery packs or fuel cell stacks. Therefore a comprehensive approach to derive an impedance model for large-scale systems is proposed in this thesis.

5.3. Impedance Model for SOFC Stacks

In this section a dynamic electrochemical model for a SOFC stack is derived step by step. Some of these steps include preceding work by other authors of the IWE. In these cases, adequate reference is given. The different steps yield very elaborate models that have their own valuable application. But an impedance model alone only reproduces the dynamics of an electrochemical system. Hence, an underlying static model has to be designed first. In section 5.3.1 the calculations, that are needed to provide the necessary information for a static micro-model, are introduced. This is expanded to a large-scale model in section 5.3.4.

In section 5.3.1 also a static micro-model is introduced that accounts for the static behavior of a small-scale system. Based on an ECM, a micro-model is introduced in section 5.3.2, with which the electrochemical losses can be calculated. The dynamics of a small-scale system can be simulated by the dynamic micro-model in section 5.3.3, which is based on the electrochemical impedance of a small-scale system. In section 5.3.4, the static micro-model from section 5.3.1 is expanded to a static macro-model. Correspondingly, the same is done with the model from section 5.3.2 to calculate the electrochemical losses for large-scale systems in section 5.3.5. Last, an attempt to derive a dynamic macro-scale impedance model is described in section 5.3.6.

---

[2]Impedance model is used here as a more general expression than ECM, because a general impedance model can be given by any complex function and is not restricted to a ECM. Nevertheless, impedance models discussed in this thesis will all be ECMs.

5.3.1. Static Micro-Model for SOFC (Behavior Model)

The model derived in this section is intended for an A_1 SOFC single cell in H_2/H_2O operation. Constant operating conditions including time invariance for the cell are assumed. The assumption holds true or causes negligible errors for the applied Anode Supported Cells (ASCs) for all relevant operating parameters as announced in the following list:

- **Temperature.** There is no evidence, that the temperature of the whole active area of the electrodes and the electrolyte can be assumed constant, but there are strong indications, that underline the fact, that no significant temperature gradient is expected neither alongside nor across the cell:

  - The waiting time after a temperature change or load change is always long enough to ensure, that the whole cell is in thermal equilibrium.

  - The heat produced by cell reactions is small compared to the heat capacity of the overstochiometric gas flow, that is used as a standard in the test bench[3].

  - It is further assumed, that the temperature rise due to increased current densities is not large. But it might have an impact on the results of the comparison of simulation and measurement for the macro-model. In this case it is assumed that the rise in temperature for both the small-scale and the large-scale system are comparable, so that the model is adequately parameterized.

- **Cathode gas.** The cathode gas used in this thesis is air with a flow rate of $Q_{gas} = 250\,\text{ml/min}$. The equivalent current density that can be maintained with the corresponding flow of O_2 molecules is $I > 7\,\text{A}$. So the gas concentration at $I = 3\,\text{A}$ would consume less than half of the O_2 in the cathode gas and would change the oxygen partial pressure from $p_{O_2,\text{inlet}} = 0.21\,\text{atm}$ to $p_{O_2,\text{outlet}} = 0.105\,\text{atm}$. Such a change has very little consequences for the electrochemical properties of the cell operation and is neglected here [39].

- **Anode gas.** It has been shown in [176], that the maximum impact of anode gas consumption in H_2/H_2O operation is $\eta_{GC} < 3\,\text{mV}$ for $Q_{gas} = 250\,\text{ml/min}$ and is therefore in the accuracy range of the proposed electrochemical model.

The static behavior is commonly analyzed via Current/Voltage (C/V) curves, which were introduced in section 2.3.2.1. It can be measured by adjusting temperature, p_{H_2O} and slowly varying the current density. A typical performance map obtained from a measurement at $T = 800\,^\circ\text{C}$, $p_{H_2O} = 0.05...0.80\,\text{atm}$ and $J = 0...2\,\text{A/cm}^2$ is shown in figure 5.1. C/V measurements have been conducted down to a minimum cell voltage of $U_{op} = 0.7\,\text{V}$ to prevent cell damage. A corresponding behavior model can easily be deduced, as the C/V curves depend almost linearly on the current density for a wide range of operating parameters. In order to demonstrate this fact, a linear plane was fitted to the performance map in figure 5.1. In figure 5.2 the deviation of measured values and linear fit are plotted over the operating conditions p_{H_2O} and J.

The C/V characteristics yield the electrochemical losses for every operating point, η_{tot}. They are given by the difference between theoretical cell voltage and measured terminal

---

[3]Some exceptions apply for the measurements reported in section 2.5.2.2.

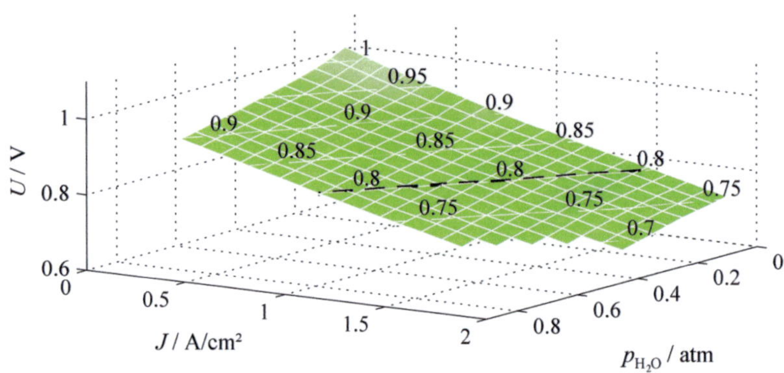

Figure 5.1.: Measured performance map of sample S.7a; the following operation parameters were chosen: $p_{H_2O} = 0.05...0.80$ atm, $J = 0...2$ A/cm^2, $T = 800\,°$C. The dashed line indicates the distributed operating conditions of the cell at $U_{op} = 0.8$ V.

voltage, as described by equation (2.28). In figure 5.3 they are plotted over the operating parameters together with the corresponding output power density P_{tot} (see equation (2.27)).

The electrochemical efficiency can be calculated by the quotient of U_{op} and U_{OCV} in every operating point. This is shown in figure 5.4. It can be seen that the best efficiency is obtained for very low current densities, however, the power density of a system in this operating point is very low. As a consequence, larger systems are needed to provide equal output power to a system operated at higher current densities. It will be demonstrated in

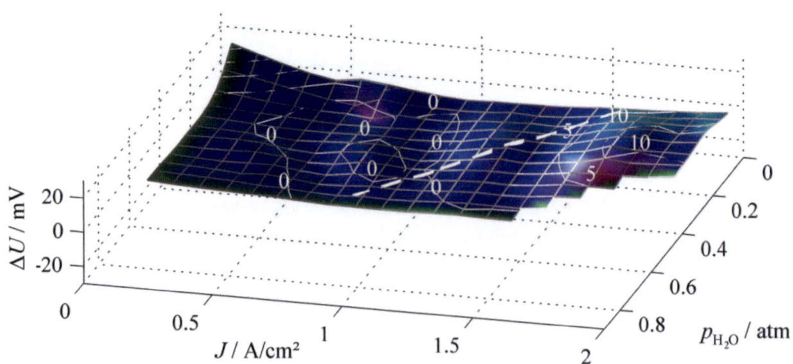

Figure 5.2.: A linear plane was fitted to the parameter map shown in figure 5.1 for $p_{H_2O} = 0.3...0.6$ atm, $J = 0.3...1.2$ A/cm^2; this figure shows the deviation of the measured values from this linear plane in mV. The dashed line indicates the distributed operating conditions of the cell at $U_{op} = 0.8$ V.

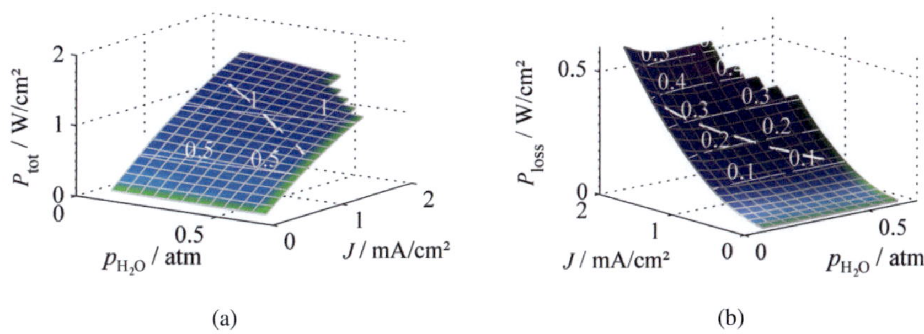

(a)   (b)

Figure 5.3.: (a) Power density calculated from the parameter map in figure 5.1 with $P_{\text{tot}} = U_{\text{op}} \cdot J$ and (b) corresponding power loss density calculated via $P_{\text{loss}} = (U_{\text{OCV}} - U_{\text{op}}) \cdot J$. The dashed line indicates the distributed operating conditions of the cell at $U_{\text{op}} = 0.8\,\text{V}$, respectively.

detail later, that the performance of a technically relevant system cannot be characterized by a single operating point.

Another critical point in this context is that a behavior model found to reproduces the C/V characteristics adequately does not provide more insight into the individual loss processes involved, but only yields the sum of all losses. In section 5.3.2 an approach, that provides this additional information, is proposed. It is based on the ECM developed by Leonide [39].

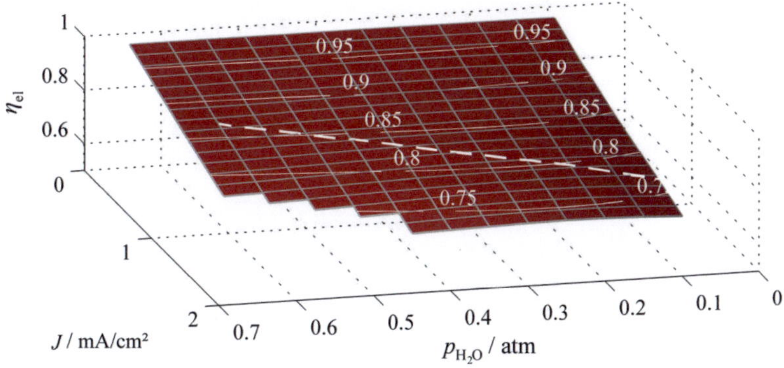

Figure 5.4.: Electrochemical efficiency determined from the measured parameter map in figure 5.1 ($\eta = U_{\text{op}}/U_{\text{OCV}}$). The dashed line indicates the distributed operating conditions of the cell at $U_{\text{op}} = 0.8\,\text{V}$.

5.3.2. Static Micro-Model for SOFC (Detailed Electrochemical Model)

The identification of electrochemical losses for SOFCs has quite a long history in SOFC research. The corresponding literature reaches from early approaches on electrolyte supported cells to advanced models for ASCs. A large variety of models with also different results can be found.

It has already been described in section 3.5.1, that EIS is a suitable technique to identify different loss contributions in one operation point. In [39] an impedance based approach to separate the loss contributions of the electrochemical processes that occur during operation of a A_1 ASC, as described in section 3.2.1 was presented. The identified processes are arranged following their characteristic frequency in table 5.1.

This model was obtained by detailed impedance analysis including a very large number of experiments with a variation of all operating parameters [39]. The model accounts for the losses of the cell operated at Open Circuit Voltage (OCV). It is assumed, that under load no further processes occur, because temperature and gas conversion effects are neglected for this micro-model, as pointed out in section 5.3.1.

Starting from this, there are two major routes to identify the overall loss contribution of the individual loss processes. They are visualized in the schematic shown in figure 5.5. The basis is the identification of loss processes in OCV, which was already conducted in [18, 39]. From there, two routes are viable to obtain the losses under load: the measurement route and the modeling route. Both will be explained in the following subsections.

5.3.2.1. Measurement Route

The measurement route is the more straightforward of the two routes, but involves a lot of fitting. The fitting can be automated by the approach presented in section 4.5.5.

	element	f_{char}, ASR	dependency	physical origin
R_0	$\mathrm{R_0}$	$> 1\,\mathrm{MHz}$, $40\ldots850\,\mathrm{m\Omega cm^2}$	T	ohmic resistance (electrolyte)
P_{1C}	RQ	$0.3\ldots10\,\mathrm{Hz}$, $2\ldots100\,\mathrm{m\Omega cm^2}$	$p_{\mathrm{O_2}}$, T (low)	gas diffusion in the pores of the cathode structure
P_{2C}	Gerischer	$2\ldots500\,\mathrm{Hz}$, $10\,\mathrm{m\Omega cm^2}\ldots2\,\mathrm{\Omega cm^2}$	$p_{\mathrm{O_2}}$, T	oxygen surface exchange kinetics and O^{2-}-diffusivity in the cathode bulk material
P_{1A}	GFLW	$4\ldots10\,\mathrm{Hz}$, $25\ldots150\,\mathrm{m\Omega cm^2}$	$p_{\mathrm{H_2O}}$, $p_{\mathrm{H_2}}$, T (low)	gas diffusion in the anode substrate
P_{2A}	RQ	$200\,\mathrm{Hz}\ldots3\,\mathrm{kHz}$, see P_{3A}	$p_{\mathrm{H_2O}}$, $p_{\mathrm{H_2}}$, T	$(\mathrm{P_{2A}} + \mathrm{P_{2A}})$ gas diffusion coupled with charge transfer
P_{3A}	RQ	$3\ldots50\,\mathrm{kHz}$, $R_{\mathrm{2A}} + R_{\mathrm{3A}}$ $= 10\,\mathrm{m\Omega cm^2}\ldots2\,\mathrm{\Omega cm^2}$	$p_{\mathrm{H_2O}}$, $p_{\mathrm{H_2}}$, T	reaction and ionic transport in the Anode Functional Layer (AFL)

Table 5.1.: Electrochemical processes identified for ASCs in [39], reproduced from [177].

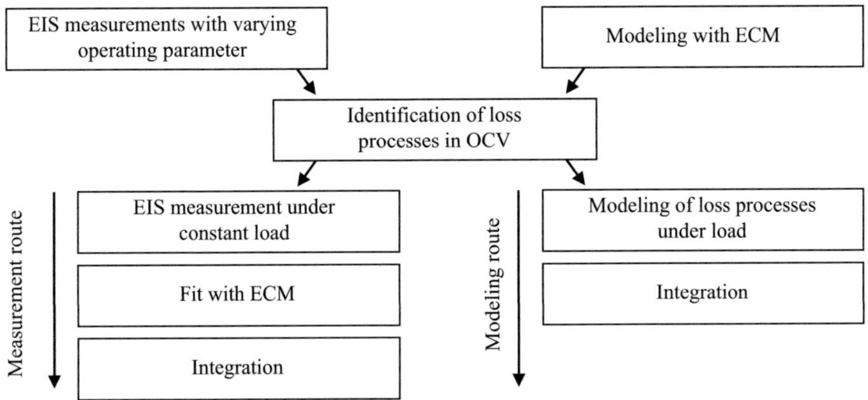

Figure 5.5.: Schematic of the two strategies to model the losses for the detailed electro-chemical model in the operating point.

EIS Measurement under Constant Load

The principle of the underlying measurement technique has been described in section 3.5.1.7. For the task here, a series of EIS has to be recorded as shown in figure 5.6. Beginning at OCV and the desired operating conditions, the current is successively increased. The corresponding current densities are $J_n = n/5 \cdot J_{\max}$ with $n = 0, \ldots, 5$.

This is necessary, because the SOFC cannot be assumed as linear system. Hence, not the ASR in the operating point but the course of the ASR has to be determined, as pointed out in section 2.3.2.1.

On the other hand, the nonlinearity in the C/V curve of the SOFC is not very distinct, so that only a relatively small number of EIS measurements is necessary to account for this nonlinearity. Depending on the desired accuracy of the obtained results, five to ten equally distributed EIS measurements are seen as adequate.

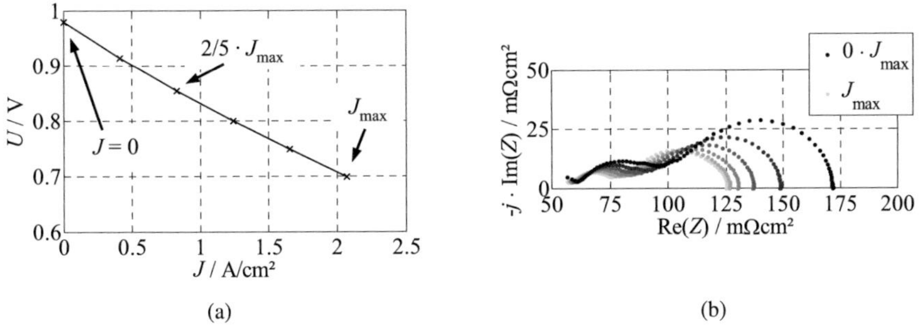

(a)   (b)

Figure 5.6.: (a) Static measurement point on the C/V curve, where EIS was conducted with sample S.8. (b) Impedance spectra measured at the measurement points indicated in figure 5.6(a).

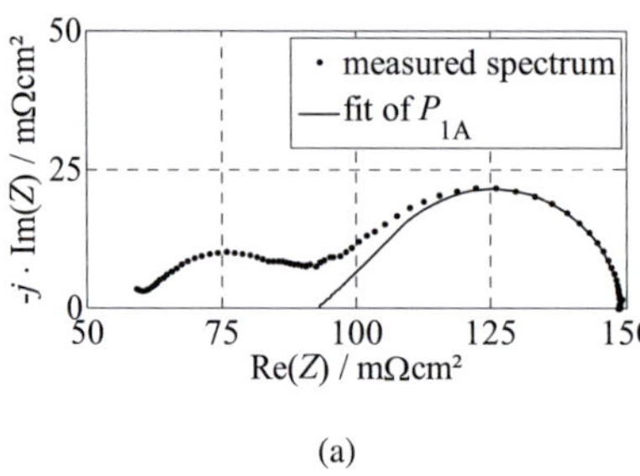

(a)

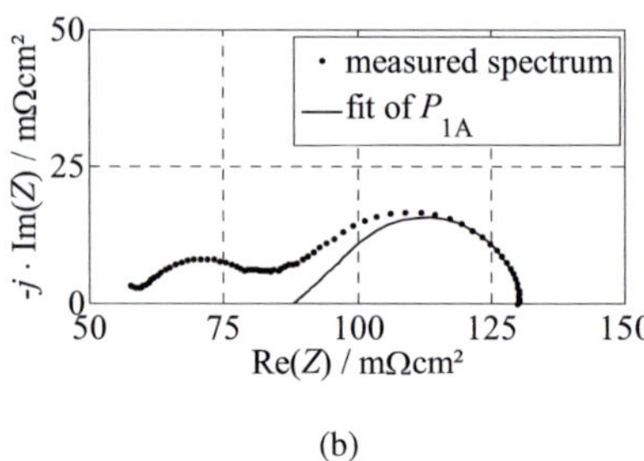

(b)

Figure 5.7.: Two exemplary spectra recorded at different current densities J (sample S.8), for which a CNLS fit was conducted, (a) for $J = 2/5 \cdot J_{\mathrm{max}}$ and (b) for $J = 4/5 \cdot J_{\mathrm{max}}$.

Fit with ECM

With the help of the impedance model developed by Leonide in [39], $Z_{\mathrm{DC},k}(J)$, the contribution of the individual processes to the overall loss (see section 2.3.2.1), is obtained for every measured EIS under load. Therefore a Complex Nonlinear Least Squares (CNLS) fit has to be conducted for every spectrum of the series depicted in figure 5.6(b).

In figure 5.7 two exemplary spectra are shown with one fitted $Z_{\mathrm{DC},k}$ also indicated. As stated before, the nonlinearity of the SOFC is not very pronounced. So a finer resolution of the function $Z_{\mathrm{DC},k}(J)$ can be achieved by polynomial fit or cubic interpolation. This is shown in figure 5.8(a).

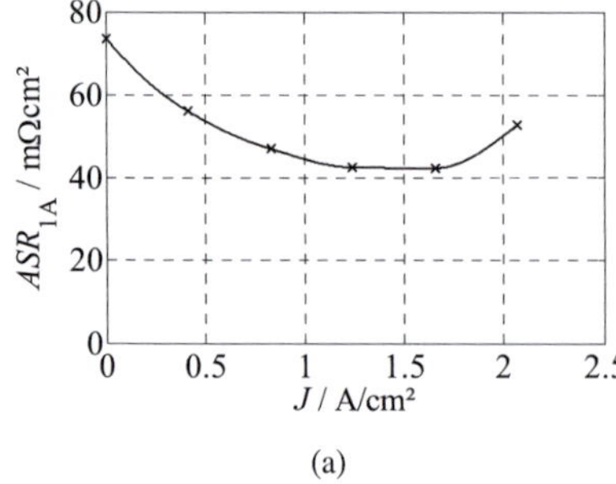

(a)

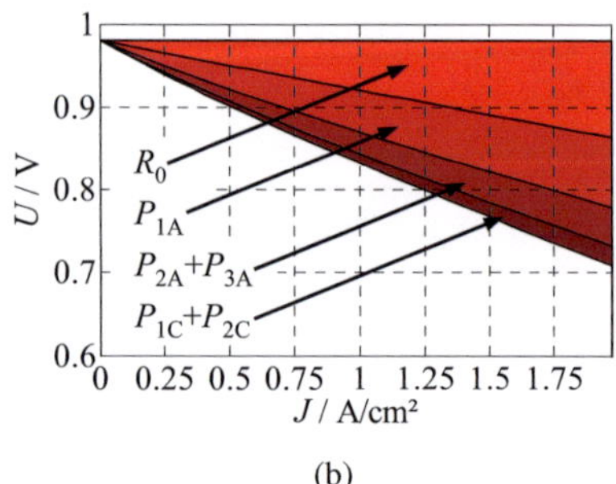

(b)

Figure 5.8.: (a) Course of $Z_{\mathrm{DC},k}(J)$ obtained as polynomial fit of the $Z_{\mathrm{DC},k}(J_n)$ determined via CNLS fit on the spectra shown in figure 5.6(b). (b) C/V plot showing the C/V curve, from which the different overpotentials η_k were subtracted consecutively.

Integration

The corresponding overpotential in the operating point is a function of the current density J_{op} as follows:

$$\eta_k(J_{\mathrm{op}}) = \int_0^{J_{\mathrm{op}}} Z_{\mathrm{DC},k}(J)\,\mathrm{d}J. \tag{5.1}$$

In most cases the values will be calculated numerically. Then the discretized form of equation (5.1) is more convenient:

$$\eta_k(J_{\mathrm{op}}) = \sum_{n=1}^{N} n\Delta J \cdot Z_{\mathrm{DC},k}\left(\frac{2n+1}{2}\Delta J\right) \quad \text{for} \quad J_{\mathrm{op}} = N\Delta J. \tag{5.2}$$

Evaluation of equation (5.1) or (5.2) yields the data necessary to calculate the power loss density with the help of equation (2.27). The power loss density will be utilized as measure for the loss processes. The motivation for this choice is explained in section 2.3.4. A more detailed description of this procedure can be found in [173, 178].

5.3.2.2. Modeling Route

The core of the modeling route is the C/V model by Leonide [104, 179]. It is an advanced approach for modeling the overpotentials of a SOFC single cell for H_2/H_2O [104] and CO/CO_2 operation [179] and is described in [18] in great detail. Here, the main characteristics will be summarized.

Modeling of Loss Processes under Load

The model introduced in [104] returns the individual overpotentials ascribable to the corresponding electrochemical processes dependent on the current density, $\eta_k(J_{\mathrm{op}})$. Hence, equation (2.27) can be evaluated easily.

The model is built upon physically based or semi-empirical approaches correlating the overpotential with the corresponding current density. The ohmic losses are modeled by an Arrhenius equation. The gas diffusion in the pores of the cathode structure and in the anode substrate is modeled by inserting Fick's law into the Nernst equation. The electrode activation overpotential was modeled by the Butler-Vollmer equation. For the exchange current density herein, a power law ansatz was applied.

The values needed to parameterize the applied equations were identified by detailed impedance analysis and microstructure analysis. Some quantities depend strongly on the cell structure, namely on the layer thicknesses, porosities and pore sizes, which are important for the diffusion processes. Others are also material-specific like the exchange current density, for example. It has to be noted, that the model works with good accuracy only for the cell, for which the parameters were identified. This becomes evident when keeping in mind, that the manufacturing tolerance for these prototype cells can reach up to 10% in performance for nominally equal cells. For more reproducible cells, the model identification has only to be conducted only once for a given exemplar. A comprehensive model, that covers all applicable materials exclusively based on physically motivated parameters, has yet to be published.

In [13] it could be shown, that the model is also applicable for Solid Oxide Electrolysis Cell (SOEC) operation. The same approach was applied to LIBs in [129].

With this model, C/V curves can be calculated for any given operating condition. It represents a well-established example of a detailed static electrochemical micro-model that is not restricted to a specific operating condition but can reproduce the static system response for every relevant operating point with good accuracy.

5.3.3. Dynamic Micro-Model for SOFC

Using the approaches in the last section, the cell voltage cannot be simulated dynamically. Only the overpotentials for a static operation point can be calculated. But the structure is not intended to include information about the dynamics of the system.

However, a dynamic micro-model is given by the impedance in the operating point. If an impedance spectrum in this particular operating point is made available through measurement, the dynamic model is easily obtained. If not, the procedure to compile the impedance of other operating points, where no spectrum was measured is not trivial but is explained in detail for the dynamic macro-model in section 5.3.6. At this point no further explanation will be provided, because the dynamic macro-model is also applicable for small-scale systems, if the geometries are chosen correspondingly.

Here some approaches found in the literature for dynamic micro-models are summarized. As already stated the dynamics of any electrochemical system in a certain operating point are described sufficiently by its impedance. However, it might be more convenient to transform the impedance into a format in which a direct time domain simulation is possible. A widely used format for this purpose is a so-called 'state-space model' [26]. There have been various attempts to design an adequate state-space model for SOFCs [180, 181]. Nevertheless, the restriction of linearity and the limited validity of a small area around the operation point make an ordinary time-invariant state-space model with a constant dynamic and output matrix unsuitable for an electrochemical system. Another critical issue is the quality of the impedance of most electrochemical systems. As stated in section 2.3.3.5, most electrode and diffusion processes cannot be modeled by easily transformable and simulatable RC circuits but necessitate RQ circuits or Transmission Line Models (TLMs). The analytical transformation into a time domain model requires fractional calculus, which was conducted in [37]. This applies for both SOFCs and LIBs.

A promising approach is the method described in [109], where the impedance of a LIB was measured at different operating points, and then transformed into a Distribution of Relaxation Times (DRT) representation which can be rearranged directly into a state-space model. If the cell's operation point lies between two of the measured nodes, an interpolation is applied to determine the current impedance. This is a very promising approach for LIBs and has the potential to build up a cell or battery behavior model, that is easy to derive and easy to apply. This approach has not been tested for SOFCs as of yet.

A physically motivated dynamic SOFC model is yet to be developed. There are several approaches for single processes like the gas diffusion in the anode substrate, which was modeled in [182, 183]. The ASR can be modeled adequately, but the characteristic frequencies vary by factors of up to 10.

5.3.4. Static Macro-Model for SOFC (Behavior Model)

As already anticipated in section 5.1.3, an expansion of the static micro-model from section 5.3.1 alongside the gas channel can account for the main difference between a small-scale and a large-scale SOFC system: the change in operating conditions alongside the gas channel. The static realization will be presented in the following subsections, starting with a record of the assumptions for this model and why they are applicable.

5.3.4.1. Assumptions

In the present version two of the three operating parameters are still assumed constant. Only the anode gas or p_{H_2O} is subject to a variation alongside the gas channel, whereas cathode gas and temperature are assumed constant. Additionally, only a current perpendicular to the electrode planes is regarded (as in [184]).

Anode Gas

The modification for the large-scale model affects only the anode gas. A sufficient parameter to account for the anode gas in H_2/H_2O operation is the steam partial pressure p_{H_2O}. In a small-scale single-cell measurement the anode gas composition and therefore p_{H_2O} is assumed to be constant. For technically relevant systems and therefore on the stack level, p_{H_2O} cannot be assumed to be constant. Commonly $p_{H_2O} \approx 0 \ldots 0.2$ atm at the gas inlet and $p_{H_2O} \approx 0.8$ atm at gas outlet are adjusted to achieve high system energy efficiency. As the fuel gas diffuses through the anode, hydrogen molecules are adsorbed and react with oxygen ions to water, increasing p_{H_2O} along the gas channel.

Cathode Gas

For the cathode gas the same applies in principle: the p_{O_2} varies along the gas channel. However, the cathode gas is ambient air which is cheap, plus the stack produces more heat than it needs to maintain operating temperature. So the air on the cathode side is commonly used to cool the stack and control the temperature. The practical gas flow rate on the cathode side is up to four times the overall gas flow rate on the anode side and thus only a little share of the oxygen in the air on the cathode side is consumed while passing the cathode gas channel. Hence, the approach presented here assumes an overstochiometric supply of air on the cathode side and a constant value for the oxygen partial pressure $p_{O_2} = 0.21$ atm.

Temperature

Different model calculations exist that predict the temperatures inside a SOFC stack [29, 76, 185, 186]. In [29] a temperature gradient of about $20\,K$ is suggested for sensible operating conditions. However, detailed measurements were not available for this study, so that the model as used in this thesis is isothermal. However, provisions for a locally varying operating temperature are available for this model (see also section 5.3.1).

Operating Voltage

The operating voltage is regarded as one single value valid for the whole cell. This means anode and cathode have only one fixed potential, respectively. This assumption is fulfilled with a very small error for high current densities because of the good electrical conductivity of both electrodes.

5.3.4.2. Modifications compared to Micro-Model

The varying p_{H_2O} along the gas channel explained in the previous section necessitates one further dimension for the model, the position x alongside the gas channel with the length l_{GC}, which is descretized in direction of the gas flow into N intervals. The operating conditions are assumed to be constant for one length element Δx:

$$p_{H_2O}(x|x_n \leq x < x_{n+1}) = p_{H_2O}(x_n), \quad n = 0, \ldots, N. \tag{5.3}$$

The distance between consecutive nodes x_n and therefore the length of each interval Δx is

$$\Delta x = x_{n+1} - x_n = \frac{l_{GC}}{N}. \tag{5.4}$$

The gas inlet and the gas outlet are defined as

$$x_{\text{inlet}} = x_0 = 0 \quad \text{and} \quad x_{\text{outlet}} = x_N = N \cdot \Delta x = l_{GC}, \tag{5.5}$$

respectively. Such a discretization was also conducted in [184].

The anode gas is given by the function $p_{H_2O}(x_n)$ and contains the different values for p_{H_2O} at every position x_n. It is worth mentioning that $p_{H_2O}(x_n)$ contains varying values for $J_{op} \neq 0$ in the static model, but is constant for all times if $J_{op} = 0$.

The function $T(x_n)$ contains the temperatures and is constant for all intervals x_n, as is the function $p_{O_2}(x_n)$, which contains the p_{O_2} on the cathode side, as explained in section 5.1.3.

The operating conditions are no longer spatially static but functions of x_n. The obtained model consists now of exactly N separate micro-models, that are arranged along x. This way a micro-model as presented in section 5.3.4 can be used to parameterize the macro-model derived here. For this, only the parameter map shown in figure 5.1 is required.

One further function has to be known to conduct a simulation of macro-model: The evolution of $p_{H_2O}(x_n)$. It can hardly be measured in an experimental setup and has to be calculated[4]. This is demonstrated in the next section.

5.3.4.3. Calculation of $p_{H_2O}(x_n)$

Two different ways to calculate the evolution of $p_{H_2O}(x_n)$ will be presented in this thesis, one is based on the differential equation for the particle conservation and the other on the equivalent current (see section 2.2.3). Both approaches are equivalent and are based on the assumption, that the conversion of every hydrogen molecule to water is accompanied by a flux of two electrons from cathode to anode. The differences of the approaches will be pointed out in the corresponding sections. The different names of the approaches are chosen in order to make them distinguishable. The operating voltage U_{op} is assumed constant over time and has to be selected prior to the model calculation for both methods.

---

[4]In [48] a test setup is described, in which the gas composition can be measured by a gas chromatograph at 5 extraction points along the gas channel. However, this method is not appropriate this experiment with H_2O as one of two gas components, as H_2O condenses on the way to the chromatograph, which works at ambient temperature (see also section 3.7.2.1).

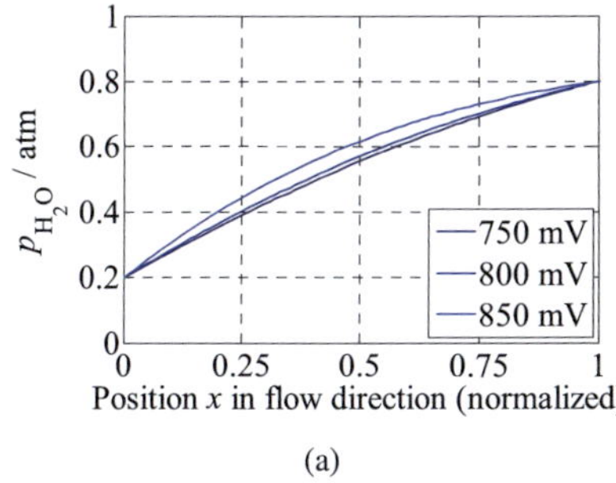
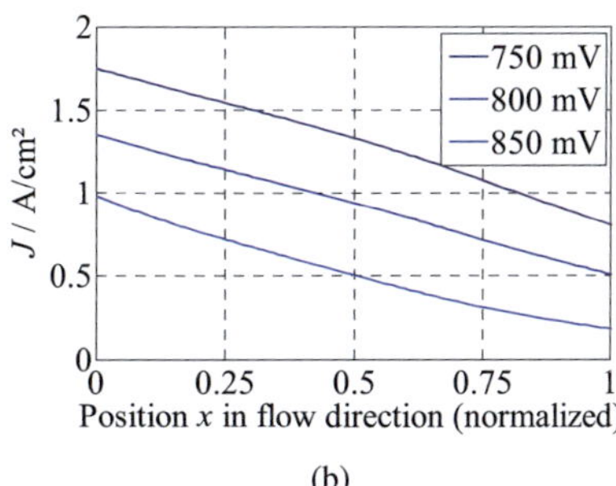

Figure 5.9.: (a) Course of p_{H_2O} for three different operating voltages calculated with the particle conservation method and (b) corresponding courses of the current density.

Particle Conservation Method

The differential equation for the conservation of particles states, that a change in the gas composition on the anode side results from a current density:

$$\frac{\Delta p_{H_2O}(x_n)}{\Delta x} = k_{gcf} \cdot J(p_{H_2O}(x_n), U_{op}). \tag{5.6}$$

Equation (5.6) links the current density in each length element Δx with the corresponding consumption of hydrogen and the corresponding production of steam. $p_{H_2O}(x_n)$ can then be calculated via different methods, as stated in [112, 178]. It should be mentioned, that the geometrical data and gas flow rates do not have to be known for this method because they are lumped in the parameter k_{gcf}, which is obtained in parallel with $p_{H_2O}(x_n)$.

With the boundary conditions $p_{H_2O}(0)$ and $p_{H_2O}(l_{GC})$, the p_{H_2O} at the gas inlet and at the gas outlet, respectively, plus the operating voltage, U_{op}, the differential equation (5.6) can be solved. The function $J(p_{H_2O}, U_{op})$ also has to be known. This can be achieved by measurement or modeling. The static micro-model introduced in section 5.3.1 or the measured performance map shown in figure 5.1 are appropriate for this task.

This method was already proposed in [112] and was performed in [173, 178] with experimental data of ASCs. The results shown here were calculated for $p_{H_2O}(0) = 0.2\,\text{atm}$, $p_{H_2O}(l_{GC})) = 0.8\,\text{atm}$ and $N = 1000$. The value for k_{gcf} has to be fitted, so that equation (5.6) is valid. $p_{H_2O}(x_n)$ can then be calculated for different operating voltages U_{op}. Note that for every operating voltage U_{op}, the value for k_{gcf} has to be readjusted because it contains information about the geometry and the gas flow in the system. The gas flow cannot be constant for equal boundary conditions and different operating voltages due to the higher gas consumption, that is a consequence of the higher current density at lower operating voltages. Figure 5.9(a) shows the calculated courses for $p_{H_2O}(x_n)$ for three different operating voltages. With this result, the calculation of the local current density is possible by simply evaluating $J(x_n) = J(p_{H_2O}(x_n), U_{op})$. The calculated courses of the local current densities for the courses of p_{H_2O}, shown in figure 5.9(a), are shown in figure 5.9(b).

The plausibility of these results can be confirmed by the following considerations: Near the gas inlet ($x = 0$), the low $p_{\mathrm{H_2O}}$ and the high fuel content in the anode gas result in a high current density. In turn, the high current density results in a large gradient in the course of $p_{\mathrm{H_2O}}$ because a large percentage of hydrogen is converted to steam at high currents. Similarly, near the gas outlet ($x = l_{\mathrm{GC}}$), only a low current density is present and the course of $p_{\mathrm{H_2O}}$ is correspondingly flat.

The results of these calculations give valuable information about the expected performance of a SOFC stack. But also they provide essential information about the state of different areas of the stack during operation. This knowledge can be utilized for the estimation of degradation rates of different areas within the cell area.

Equivalent Current Method

The second method for the calculation of $p_{\mathrm{H_2O}}(x_n)$ is based on the same principle of the conservation of particles, but has different boundary conditions and model input values. If the gas flow rate Q_{gas} of the fuel gas and its $p_{\mathrm{H_2O}}$ are known, the operating voltage determines the course of $p_{\mathrm{H_2O}}$ including $p_{\mathrm{H_2O}}(x_N)$ by the function $J(p_{\mathrm{H_2O}}, U_{\mathrm{op}})$.

This method first calculates the equivalent current I_{eq} of the provided fuel gas with equation (2.3):

$$I_{\mathrm{eq}}(0) = \frac{\left(1 - \frac{p_{\mathrm{H_2O}}(0)}{p_0}\right) \cdot Q_{\mathrm{gas}}}{V_{\mathrm{m}}} \cdot z \cdot F. \tag{5.7}$$

Then the current density in the first length element is determined by evaluating $J(p_{\mathrm{H_2O}}, U_{\mathrm{op}})$, and subsequently subtracted from the equivalent current:

$$I_{\mathrm{eq}}(x_{n+1}) = I_{\mathrm{eq}}(x_n) - A(\Delta x) \cdot J(x_n). \tag{5.8}$$

This results in a higher $p_{\mathrm{H_2O}}$ for the second length element:

$$p_{\mathrm{H_2O}}(x_{n+1}) = 1 - \frac{V_{\mathrm{m}} \cdot I_{\mathrm{eq}}(x_{n+1})}{z \cdot F \cdot Q_{\mathrm{gas}}}, \tag{5.9}$$

because the gas volume remains unchanged by the anodic reaction. This way, it is possible to calculate $p_{\mathrm{H_2O}}(x_n)$ stepwise.

The ability of this approach was analyzed systematically by a detailed validation. This validation is presented in section 5.3.4.4.

5.3.4.4. Validation of the Static Macro-Model

This validation involves experiments on different cell setups and modeling and is therefore divided into several parts.

Motivation

The aim of this validation is to show, that the modeling approach of the static macro-model, that was introduced in this section, is applicable. This means, it has to be shown that the micro-model from section 5.3.1 can be expanded to a macro-model.

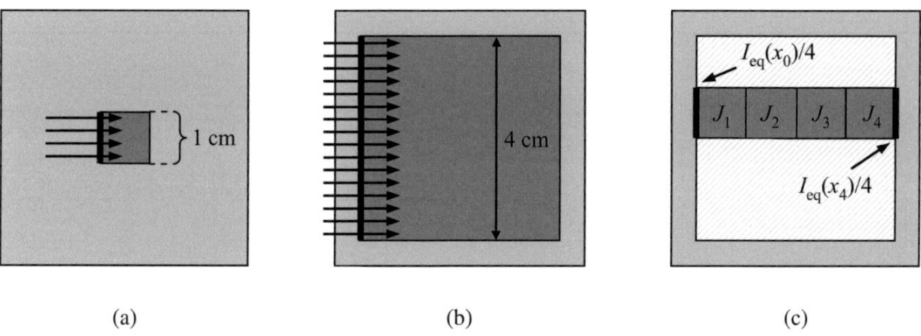

Figure 5.10.: Schematics of (a) a A_1 cell and (b) a A_{16} cell demonstrating the equal gas flow rates per flow width for model $\mathcal{A}$. (c) Scheme of the discretization of the A_{16} cell with 16 small-scale systems.

For this task a large-scale system has to be measured and simulated. As already stated, an A_{16} cell is chosen to represent a large-scale system in this thesis. The macro-model to simulate this large-scale system has to be parameterized by measurement results of a small-scale system. For this purpose an A_1 cell of identical manufacturing parameters (except the size of the working cathode) was chosen.

From the two approaches for the calculation of $p_{\text{H}_2\text{O}}$, the equivalent current approach is chosen for this validation, because for this approach only the gas flow, $p_{\text{H}_2\text{O}}$ at the gas inlet and the operating voltage have to be adjusted. As the output $p_{\text{H}_2\text{O}}$ is just a result of the model and does not have to be adjusted, the recording of a measurement series is simplified.

The gas flow rates for both cells are the crucial parameters for the experiments and result from different preconditions. Basically the large-scale system should show a large gradient in $p_{\text{H}_2\text{O}}$ from gas inlet to gas outlet. Only then is it justified to apply a macro-model. On the other hand, if the flow rate is chosen too low, significant measurement noise as a consequence of an inhomogeneous gas distribution along and among all gas channels arises and the results are useless.

This is discussed in the next section.

Gas Flow Rates

Generally, a total flow rate of $Q_{\text{gas},A_1} = 62.5\,\text{ml/min}$ for the A_1 cell[5] and $Q_{\text{gas},A_{16}} = 250\,\text{ml/min}$ for the A_{16} cell[6] are seen as minimum flow rates for both setups. At these flow rates at high current densities, additional noise in the static measurement can already be noticed.

The flow rate of $Q_{\text{gas},A_1} = 250\,\text{ml/min}$ for the A_1 cell has proven to be sufficient to assume homogeneous conditions over the whole cell area and therefore an even higher flow rate is not required for this cell setup.

---

[5]$Q_{\text{gas},A_1}/l_{A_1} = 62.5\,\text{ml/min}$ for $l_{A_1} = 1\,\text{cm}$ of flow width, see figure 5.10(a)
[6]also $Q_{\text{gas},A_{16}}/l_{A_{16}} = 62.5\,\text{ml/min}$ per cm of flow width, because $l_{A_{16}} = 4\,\text{cm}$, see figure 5.10(b)

For the A_{16} cell, gas flow rates up to $Q_{\text{gas},A_{16}} = 2000$ ml/min are applicable and reasonable (depending on the experiment). But for the measurements here, the gas flow for the A_{16} cell was varied between $Q_{\text{gas},A_{16}} = 250\ldots1000$ ml/min.

Again, it should be noted that a detailed description of the corresponding measurement setups can be found in [113] for the A_1 cells and in [48] for the A_{16} cells.

Measurement A

The cell tested in measurement A was an A_1 ASC in H_2/H_2O operation. Under static conditions the performance map $J_A(p_{H_2O}, U_{\text{op}})$ was measured. The total gas flow was adjusted to $Q_{\text{gas},A_1} = 62.5$ ml/min at both anode and cathode side at all times. This means, that the gas flow is so low that the gas conversion over the cell area cannot be neglected and the operating conditions cannot be assumed homogeneous over the whole cell area. Hence, measurement A does not meet the requirements of a small-scale system. However, its purpose will become clear, when the results of model $\mathcal{A}$ are discussed.

The p_{H_2O} was adjusted to $p_{H_2O} = 0.05\ldots0.8$ atm and the current density to $J = 0\ldots3$ A/cm^2. No measurement was conducted below an operating voltage of $U_{\text{op}} = 0.7$ V to guarantee negligible degradation throughout the whole measurement.

Measurement B

In measurement B the same cell was tested as in measurement A and the measurement purpose was the same: the parameter map $J_B(p_{H_2O}, U_{\text{op}})$. The difference in the two measurements was that in measurement B the total gas flow was changed to $Q_{\text{gas},A_1} = 250$ ml/min. As stated before, former experiments had shown that this gas flow is sufficient, that the gas conversion over the cell area can be neglected and homogeneous conditions can be assumed. In [176] it was shown, that only a minor error arises from the gas conversion in this test setup with $Q_{\text{gas},A_1} = 250$ ml/min gas flow. The operation parameters (p_{H_2O}, J, U_{op}) were chosen to be the same as measurement A. Measurement B can be viewed as the measurement of a small-scale system.

Measurement C

Measurement C was conducted on an A_{16} cell in H_2/H_2O operation. Its purpose is to provide measurement results of a large-scale system. The different experiments are listed in table 5.2. It should again be emphasized, that the primary purpose of the static macro-model is to reproduce the static behavior of a SOFC stack. In this context it is most important, that the macro-model is able to simulate measurements C.1 and C.2, the ones with the highest impact of the gas conversion along the gas channel and therefore the largest variation of operating parameters in direction of x. Measurements C.3 to C.6 were conducted to further test the capability of the macro-model to reproduce also other gas flow rates.

Model $\mathcal{A}$: Modeling Measurement C with Measurement A

The first approach is a very simple form of a macro-model. The static behavior of the A_{16} cell shall be reproduced by measurements of an A_1 cell. Therefore the length of the

measurement number	anode gas flow rate Q_{gas}	$p_{\text{H}_2\text{O}}$ in atm
C.1a	250	0.200
C.1b	250	0.200
C.2	250	0.450
C.3	500	0.200
C.4	500	0.450
C.5	1000	0.200
C.6	1000	0.450

Table 5.2.: Measurements conducted for measurement C (sample S.7b).

gas channel of the A_{16} cell is discretized into only four elements ($N = 4$), so that the large-scale system can be represented exactly by 16 small-scale systems, as indicated in figure 5.10(c). This way the discetization elements will show the same behavior as the A_1 cell, as long as the same gas flow rate per cm flow width is adjusted. This is the case for measurements C.1 and C.2. The gas conversion in measurement A is not considered, because the discretization elements of the large-scale system are supposed to show the same gas conversion. The results are obtained by evaluating equations (5.8) and (5.9) iteratively, choosing $N = 4$. The current in every discretization element is calculated and subtracted from the equivalent current I_{eq} of the fuel gas.

Since not every operation point of each element is directly represented by the measured values for $J_A(p_{\text{H}_2\text{O}}, U_{\text{op}})$, a simple two-dimensional interpolation is applied. This is a viable strategy, because the system shows only very small non-linearities in the measured region.

A simple algorithm, that executes the mathematical operations described here within a loop, yields $p_{\text{H}_2\text{O}}$ before and after every length element and the current in each element.

Results Model A. Measurements and model calculations of A are compared in table 5.3. Figure 5.11 further elucidates the principle of the model. For three examples the calcu-

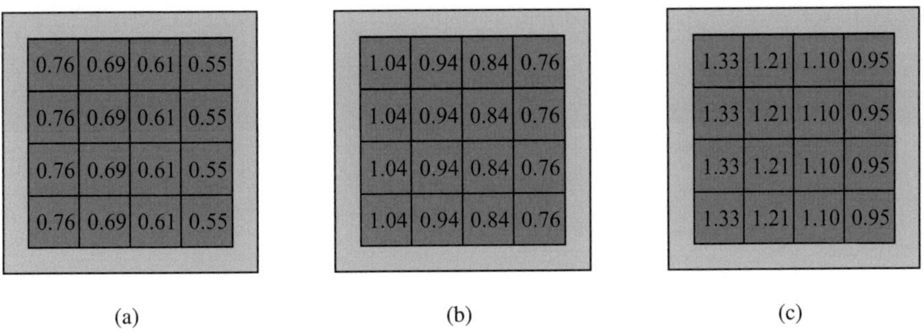

(a)   (b)   (c)

Figure 5.11.: Visualization of the results of model A, where the current in each individual element is plotted in the corresponding element for (a) C.1.3, (b) C.1.4 and (c) C.1.5.

number	U_{op} in V	I_{op} in A		$p_{\mathrm{H_2O}}(x_N)$ in atm		error in %	
		measured	modeled	measured	modeled	in I_{op}	in $p_{\mathrm{H_2O}}$
C.1.1	0.95	3.14	3.26	0.288	0.291	3.8	1.0
C.1.2	0.90	6.60	6.70	0.384	0.387	1.5	0.8
C.1.3	0.85	10.39	10.45	0.490	0.492	0.6	0.4
C.1.4	0.80	14.45	14.32	0.603	0.600	0.9	0.5
C.1.5	0.75	18.20	18.39	0.708	0.713	1.0	0.7

Table 5.3.: Results of model $\mathcal{A}$.

lated currents in the individual elements are visualized. In table 5.3, the gas compositions at the gas outlet $p_{\mathrm{H_2O}}(x_N)$ obtained from the model and calculated from the measurement are compared. The latter is calculated by subtracting the measured current on the A_{16} cell from the equivalent current supplied assuming $\eta_{\mathrm{F}} = 1$. This can also be done via equation (5.9). The simulated $p_{\mathrm{H_2O}}(x_N)$ is the result the macro-model calculates for the gas outlet. Except for C.1.1 all values show very little deviation. The calculated errors of C.1.2 to C.1.5 are considerably below 2%. This proves the applicability of the model. It further shows that edge effects, temperature effects, a non-uniform gas distribution and the influence of the cathode gas do not play a significant role in this kind of measurement.

Model $\mathcal{B}$: Modeling Measurement C with Measurement B

Measurement B was conducted with a high gas flow rate. As stated above, the operating conditions can assumed to be equal throughout the whole active cell area. This in turn means that the obtained parameter map $J_{\mathrm{B}}(p_{\mathrm{H_2O}}, U_{\mathrm{op}})$ is independent from the gas flow rate because the gas flow rate does not determine the operating conditions.

The difference between the simulations with model $\mathcal{B}$ and model $\mathcal{A}$ are that measurement B is used to parameterize model $\mathcal{B}$ and that the number of discretization elements is not fixed to $N = 4$.

Results Model $\mathcal{B}$. The number of elements N must be large for model $\mathcal{B}$. In fact, the simulations show, that the higher the number of elements, the higher the accuracy of the model. This is exemplified in table 5.4.

number of elements N	I_{op} in A	$p_{\mathrm{H_2O}}(x_N)$ in atm	error in % in I_{op}	in $p_{\mathrm{H_2O}}$
4	15.64	0.636	8.2	5.5
10	15.20	0.624	5.2	3.5
100	14.94	0.617	3.4	2.3
1000	14.92	0.616	3.3	2.2
5000	14.92	0.616	3.3	2.2

Table 5.4.: Results of model $\mathcal{B}$ for experiment C.1.4 from table 5.3 indicating the need of small discretization elements.

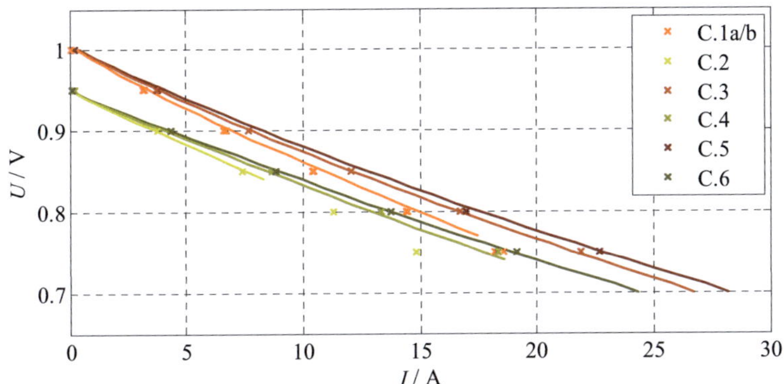

Figure 5.12.: Comparison of measured C/V curves (sample S.7b) and C/V curves simulated by the static macro-model parameterized with the values measured with sample S.7a.

The result is not surprising, because with a rising number of elements, the length of each element decreases and the gas conversion has less influence in each discretization element.

As can be seen for an increasing number of discretization elements, the output values of the model converge to values with little deviation to the ones from measurement C. It has to be emphasized again, that the gas flow rates per cm flow width of the parameterizing measurement and of the simulated measurement vary by a factor of 4.

Any flow rate can be simulated with model B by adjusting the factor Q_{gas} in equations (5.8) and (5.9). Figure 5.12 compares the values from measurement C with the corresponding simulations by model B for $N = 1000$ for C/V curves with different gas flow rates and different p_{H_2O} at the gas inlet. Again the deviations are very small and demonstrate the applicability of the proposed model to simulate any gas flow of the A_{16} cell appropriately.

Discussion

In this section the static macro-model proposed in this thesis has been validated for two cases:

- **Case 1.** Low flow rate in the small-scale system, number of descretization elements chosen, so that the simulated elements have the same size as the small-scale system.

- **Case 2.** High flow rate in the small-scale system, number of descretization elements large, so that the simulated elements are very small and the gas conversion in every element of the large-scale system is negligible.

It has already been shown above that the deviations of model B grow larger for a small number of elements. It should also be noted that the simulation results of model A for a large number of elements grows larger as shown in table 5.5. This can also be explained by gas conversion in the system. As stated before, the gas conversion cannot be neglected for

number of elements N	I_{op} in A	$p_{\mathrm{H_2O}}(x_N)$ in atm	error in % in I_{op}	error in % in $p_{\mathrm{H_2O}}$
4	14.32	0.600	0.9	0.5
10	13.95	0.589	3.5	2.3
100	13.74	0.584	4.9	3.2
1000	13.72	0.583	5.1	3.3
5000	13.72	0.583	5.1	3.3

Table 5.5.: Results of model $\mathcal{A}$ for experiment C.1.4 from table 5.3 indicating the necessity equal sizes for measured cells and modeled discretization elements for low gas flow rates.

measurement A. Therefore it is already included in the simulation of one descretization element. Consequently, gas conversion has to be the same in the measured and in the simulated element. A scaling of the element size and the corresponding current and gas conversion is only possible if the measurement of the small-scale system shows no or only negligible influence to gas conversion like measurement B does. In turn, if the small-scale system measured shows no influence to gas conversion, it is scalable, but the simulated element must be so small that it is also not affected by gas conversion. Therefore, a simulation of large elements based on measurement B yield large deviations.

In conclusion it can be stated, that there are two possibilities to simulate a large-scale system with the presented approach:

- If the gas flow rate of a small-scale system and a large-scale system are equal, the requirements on homogeneity on the small-scale system are not strict. Only the size of the measured small-scale system and the simulated element must be equal.

- If the gas flow rate of a small-scale system and a large-scale system are not equal, then the measurement on the small-scale system must strictly fulfil the homogeneity condition. The descretization elements of the large-scale system must be as small as possible to avoid influence of the inhomogeneity of the large-scale system on the simulation.

As a final remark it should be stated, that the calculation time for every simulation shown in this section was below $10\,\mathrm{sec}$ and therefore reasonably small. Hence, the simulation time is no issue for the static macro-model.

5.3.5. Static Macro-Model for SOFC (Detailed Electrochemical Model)

The identification of the electrochemical losses for large-scale systems is conducted analogue to the static electrochemical model derived in section 5.3.4. The discretization of the gas channel yields the losses per length element, as described in section 5.3.2. Figure 5.13 shows a schematic of the resulting model which is described in detail in [173]. Every element of the resulting 1D model is assigned an ECM. With the parallel connection, the assumption of constant cell voltage along the cell is implied as done before for the static macro-model in section 5.3.4.

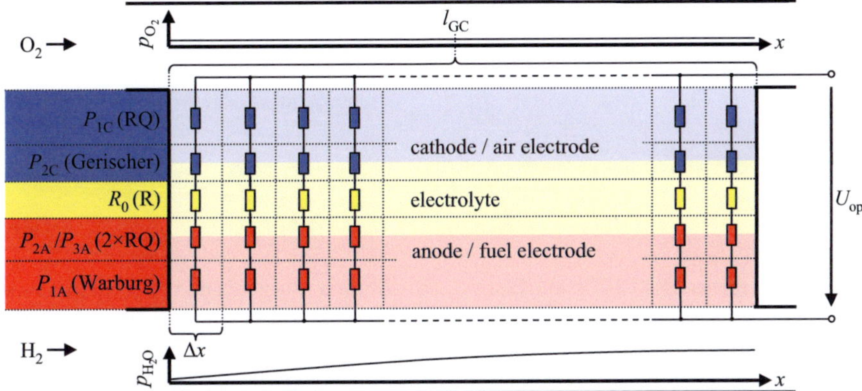

Figure 5.13.: Schematic of the 1D model discretized along the gas channel with an ECM for every element. In the gas channels on anode and cathode side the course of the partial pressures of the applied gases are shown schematically.

In order to parameterize the model, a lot of data is necessary. In addition to the parameter map, $J(p_{H_2O}, U_{op})$, also $Z_{DC,k}(p_{H_2O}, U_{op})$ has to be known. For the model in section 5.3.2, $Z_{DC,k}$ was required as function of J for the applied p_{H_2O}. For the detailed electrochemical model in this section, the whole operating area has to be covered. Therefore, a large number of impedance spectra has to be recorded and subsequently fitted. This will not be amplified here but is explained in more detail for the dynamic macro-model in section 5.3.6. However, two examples for parameter maps $Z_{DC,k}(p_{H_2O}, U_{op})$ are shown in figure 5.14. $p_{H_2O}(x_n)$ is calculated as in section 5.3.4 with the conservation of particles method. The results for different operating voltages are depicted in figure 5.9(b). The losses η_k due to process k in every descretization element are calculated analogously to

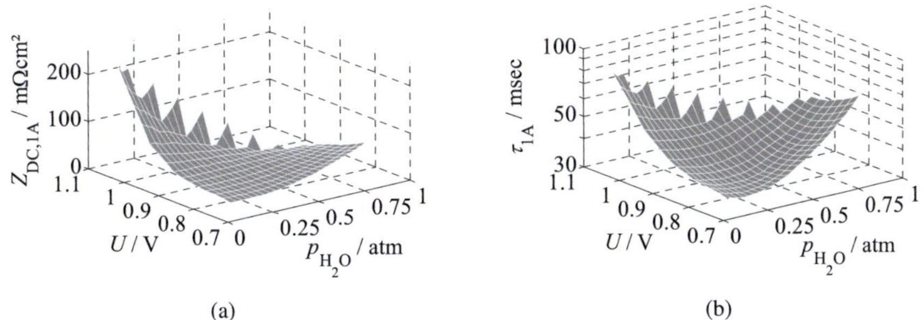

Figure 5.14.: Two examples for parameter maps $Z_{DC,k}(p_{H_2O}, U_{op})$: (a) $Z_{DC,1}$ accounting for the losses due to the gas diffusion in the anode substrate P_{1A} and (b) the corresponding time constant (sample S.8).

equation (5.1) in the continuous form via

$$\eta_k(x_n, p_{\mathrm{H_2O}}, U_{\mathrm{op}}) = \int_0^{J(p_{\mathrm{H_2O}},U_{\mathrm{op}})} Z_{\mathrm{DC},k}(p_{\mathrm{H_2O}}, J)) \, \mathrm{d}J. \tag{5.10}$$

Then the power loss density at every position x_n is calculated by:

$$P_k(x_n) = J(p_{\mathrm{H_2O}}(x_n), U_{\mathrm{op}}) \cdot \eta_k(x_n). \tag{5.11}$$

This way, the power loss can be attributed to the different physical and chemical processes in the large-scale system. Figure 5.15 shows four different power loss densities. It is obvious from the diagrams, that the major losses are generated by anode and electrolyte. This is not surprising, when comparing these results with the results of the micro-model.

It should also be noted that this calculation represents a realistic operating point for a SOFC stack in H_2/H_2O operation. But the model not yet includes additional losses from cell contacting, as ideal contacting was applied as explained in section 2.5.1.1. With this model, also different characteristics of the whole setup can be deduced. They are compiled in table 5.6. Once again, the findings of section 5.3.1 (figures 5.3 and 5.4) are confirmed: a higher power output density of a given system will lead to a lower electrical efficiency, η_{el}.

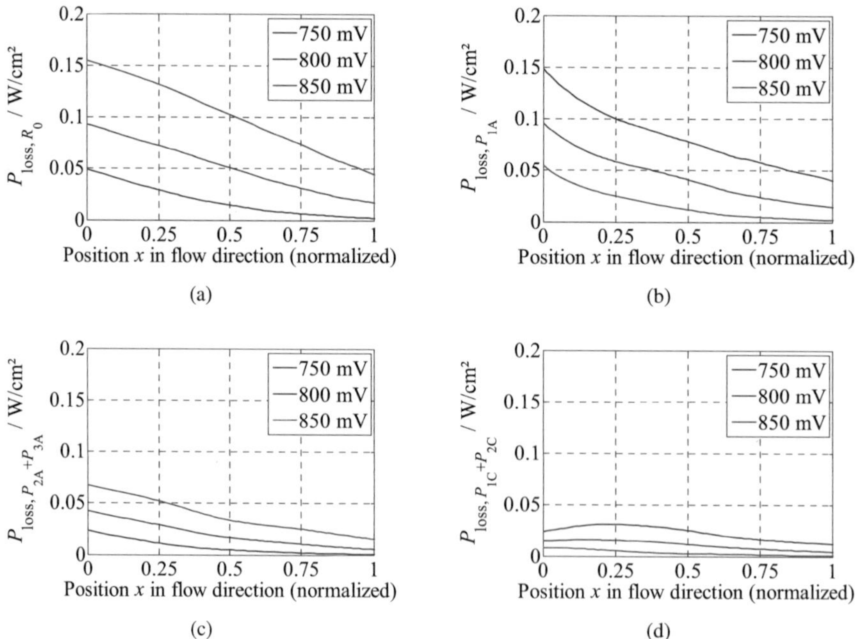

Figure 5.15.: Calculated power loss density distributions for the circuit elements from table 5.1 at different operating voltages ($T = 800\,°\mathrm{C}$) for: (a) R_0, (b) $P_{1\mathrm{A}}$, (c) $P_{2\mathrm{A}} + P_{3\mathrm{A}}$, (d) $P_{1\mathrm{C}} + P_{2\mathrm{C}}$.

quantity	calculated values for U_{op}		
	750 mV	800 mV	850 mV
mean power density in W/cm^2	979	697	393
mean power loss density (anode) in W/cm^2	119	63	24
mean power loss density (electrolyte) in W/cm^2	102	52	19
mean power loss density (cathode) in W/cm^2	23	11	4
mean power loss density in W/cm^2	244	127	46
electrical efficiency η_{el} in %	80.0	84.6	89.5

Table 5.6.: List of additional characteristics of the large-scale system that can be deduced from the macro-model.

5.3.6. Dynamic Macro-Model for SOFC

This section describes the attempt to derive a dynamic macro-model for SOFC representing the dynamic behavior of a SOFC system in a specific operation point. This operation point is defined by the static macro-model from section 5.3.4. The second approach for calculating $p_{\text{H}_2\text{O}}(x_n)$ is applied from section 5.3.4.3. Hence, the required input parameters for the model are U_{op}, $p_{\text{H}_2\text{O}}$ at the gas inlet, and the total anodic flow rate.

Like the dynamic micro-model in section 5.3.3, the dynamic macro-model is based on the impedance in the operating point. For the parameterization, the impedance of the underlying small-scale system has to be measured in every local operating point that is likely to occur in the large-scale system. The covered range of $p_{\text{H}_2\text{O}}$ and U_{op} is equal to the measurement range applied in section 5.3.4.4. In order to guarantee good accuracy, the measured nodes should be as close as possible. For the example shown here, 42 impedance spectra were recorded. Together with the settling time between EIS measurements, the total time for the experiment is approximately three days. The procedure to obtain the operation points for the sensible EIS measurements was chosen as follows: After adjusting the $p_{\text{H}_2\text{O}}$ (7 nodes covering $p_{\text{H}_2\text{O}} = 0.05\ldots0.70$ atm) and a settling time

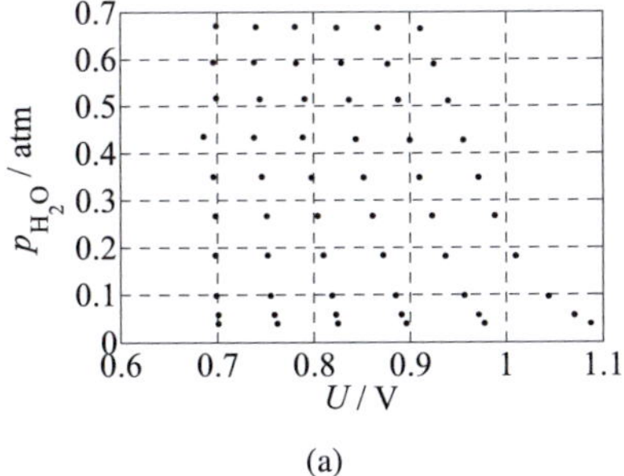
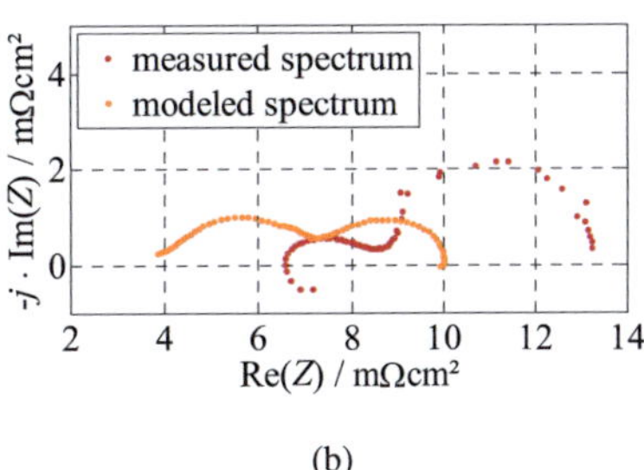

(a) (b)

Figure 5.16.: (a) Operating conditions at which EIS was conducted and (b) impedance spectrum simulated with the dynamic macro-model for $Q_{\text{gas}} = 250$ ml/min, $U_{\text{op}} = 0.8$ V, $p_{\text{H}_2\text{O}}(0) = 0.2$ atm (sample S.7a).

of $t = 30\,\text{min}$, the current density was increased in small steps, until the operating voltage reached $U_\text{op} = 0.7\,\text{V}$. The corresponding current density was defined as J_max. The EIS measurements were conducted with offset current densities of $J = n/5 \cdot J_\text{max}$ with $n = 0, \ldots, 5$ after a settling time of $t = 30\,\text{min}$. All nodes are depicted in figure 5.16(a). For every node, one complete impedance spectrum is then available, consisting of real and imaginary part. The resulting parameter maps are:

$$Z_\text{Re}(p_{\text{H}_2\text{O}}, U_\text{op}, \omega) = \text{Re}(Z(p_{\text{H}_2\text{O}}, U_\text{op}, \omega)), \tag{5.12}$$

$$Z_\text{Im}(p_{\text{H}_2\text{O}}, U_\text{op}, \omega) = \text{Im}(Z(p_{\text{H}_2\text{O}}, U_\text{op}, \omega)). \tag{5.13}$$

The course of $p_{\text{H}_2\text{O}}(x_n)$ is taken from the static macro-model from section 5.3.4 and the offset voltage is fixed to U_op. Analogue to the calculation of the current density in section 5.3.4, the impedance is calculated for every branch of the model shown in figure 5.13. Real and imaginary part of the impedance are determined separately, in order to avoid a complex parameter map for interpolation, if the values for $p_{\text{H}_2\text{O}}(x_n)$ and U_op are not represented by a node.

The impedance for one branch in obtained as

$$Z(x_n, \omega) = Z_\text{Re}(p_{\text{H}_2\text{O}}(x_n), U_\text{op}, \omega) + j \cdot Z_\text{Im}(p_{\text{H}_2\text{O}}(x_n), U_\text{op}, \omega). \tag{5.14}$$

The impedance of the whole system $Z_\text{MM}(\omega)$ is obtained as parallel connection of these branches:

$$Z_\text{MM}(\omega) = \frac{A_{16}}{l_{A_{16}} \cdot N} \left(\sum_{n=1}^{N} \frac{1}{Z(x_n, \omega)} \right)^{-1}. \tag{5.15}$$

The validation of the dynamic macro-model with corresponding EIS measurements on A_{16} cells reveals that the two spectra show little similarity as can be seen in figure 5.16(b). However, the corresponding DRTs shown in figure 5.20(c) reveal, that at least the number of processes and the corresponding frequencies are in agreement – apart from a very large low frequent process in the DRT of the measurement. In fact, the deviations can be broken down to major two shortcomings:

- The high frequent part of the measured spectrum becomes inductive very early. This is due to the low measured impedance values and the measurement setup for the A_{16} cells.

- The low frequent part of the simulated spectrum shows a completely different behavior as the measured spectrum.

Whereas the first shortcoming could not be eliminated in this thesis, the large deviation in the low frequent part of the spectrum will be discussed and an explanation will be proposed: it is due to the so-called GCI, as will be discussed in section 5.3.7.

5.3.7. Extension of the Dynamic Macro-Model: Gas Conversion Impedance (GCI)

The GCI plays a special role in a large-scale SOFC system. A simple up-scaling of the impedance in analogy with the static macro-model does not result in the measured impedance spectrum of the large-scale system as confirmed in section 5.3.6. When comparing measured and simulated data in figure 5.16(b), it becomes apparent that the impedance calculated in this way does not match the measured impedance. One reason for this is the GCI that was not considered so far but will be derived in this section. It has to be considered for SOFC systems in which the consumption of the fuel or oxidation gas has a significant influence on the system behavior and cannot be neglected.

The deviation of model and measurement shown in section 5.3.6 originates from the fact that the impedance was calculated assuming a constant course of $p_{H_2O}(x_n)$ in time. This assumption does not hold for low frequencies. For ways to overcome this problem, the phenomenological background of the GCI has to be explained. This is done in section 5.3.7.3. On this basis, a way to include the GCI in the model is proposed in section 5.3.7.4. But first of all, a brief literature survey on the GCI is presented in section 5.3.7.1, followed by an interpretation of the GCI in section 5.3.7.2.

5.3.7.1. GCI in the Literature

The GCI is discussed in multiple studies in the literature. At first, there are different names for this effect. It is called 'gas concentration impedance' in [172], 'gas transport impedance' in [187], 'gas conversion impedance (GCI)' in [188, 189]. The latter term is seen as the most accurate and will be used in this thesis. Even though it can be regarded as an effect of the gas transport and the gas concentration, it only occurs, if eventually gas is converted and if the conversion itself has an influence on the system behavior.

In [172] a very elaborate approach to determine the GCI is presented. It incorporates the Navier-Strokes equations in order to simulate the gas flow kinetics along the gas channel. For the local resistance of the anode, a Butler-Volmer approach is applied. This is a distinct advancement compared to other stack models that assume the ASR as constant throughout the whole cell area as done in [175]. Nevertheless, the impedance of the SOFC anode has not only one but two major contributions [162]. It could be shown in [39, 104], that the anode resistance can be modeled accurately by a Butler-Volmer approach plus the gas diffusion resistance (see also section 5.3.2), the latter being neglected in [172]. Furthermore, the resistances of electrolyte and cathode are of importance in order to calculate reasonable values for the local current density. Last, the study lacks a direct comparison with measured data and can only be confirmed qualitatively by comparing its findings with literature data.

In [189] an experimental setup for button cells is presented which enables the determination of the GCI. Moreover, this impedance is modeled by a Continuously Stirred Tank Reactor (CSTR) and compared to measured values. In [190] measurements on cells with an active area of $A = 16\,cm^2$ with different anode gases are shown. It is tried to ascribe the total low frequent losses to Knudsen diffusion, gas phase diffusion and gas conversion. But only primary modeling results are shown.

In [188] the GCI was measured and modeled for disc type cells with an active area of $A = 113\,cm^2$.

In [187] gas conversion and gas diffusion impedances have been identified for segmented-in-series tubular SOFCs.

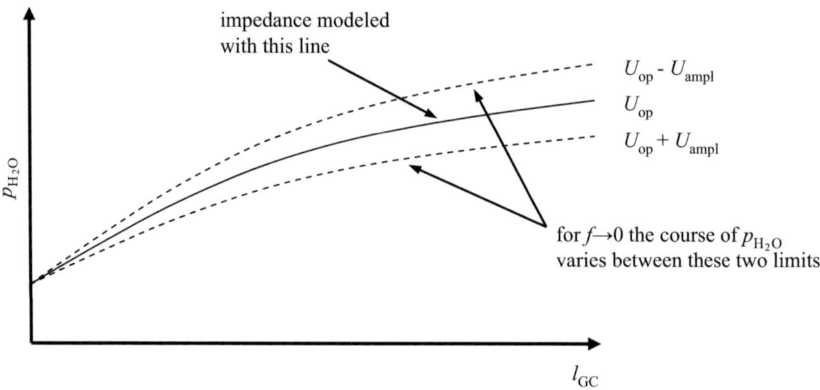

Figure 5.17.: Scheme of the course of $p_{\mathrm{H_2O}}(x_n)$ at the static operating point and when superposed with an amplitude (static case).

5.3.7.2. Interpretation of the GCI

The interpretation and categorization of the GCI should be considered briefly at this point. Even though it appears as an extra loss process in the impedance spectrum and in the DRT, it is clearly not a loss process. Hence, the voltage drop it is responsible for should not be seen as electrochemical loss. This can be explained by the fact, that a large gas conversion rate raises the fuel utilization of the system, which represents one important efficiency factor for a SOFC system. This means, a large GCI indicates a desired fuel utilization.

5.3.7.3. Phenomenological Explanation of the GCI

The gas conversion related voltage drop is correlated to the drop in the Nernst voltage (see equation 2.12) due to a depletion of fuel gas, because fuel gas has been consumed or converted in one of the prior length elements in the gas channel and is no longer available for subsequent ones.

The GCI arises because a change in the operating voltage not only results in a direct change of the local current density, $J(x_n)$, that can be calculated by the electrochemical impedance of the system as modeled and measured for small-scale systems. As a consequence of a change in the local current density, $J(x_n)$, of necessity the course of $p_{\mathrm{H_2O}}(x_n)$ is also changed. In turn, the local current density not only depends on the operating voltage but also and even distinctively on $p_{\mathrm{H_2O}}(x_n)$, as can be seen from impedance spectra in [162], for example. Hence, it becomes apparent that there is a complex interdependence of operating voltage U_{op}, the local current density $J(x_n)$, and the course of $p_{\mathrm{H_2O}}(x_n)$. This interdependence is specified by the GCI.

In other words, due to an excitation voltage, U_{ampl}, the course of $p_{\mathrm{H_2O}}(x_n)$ changes, as demonstrated in the scheme in figure 5.17. This change is observable as additional process in the impedance of the system. The effect of the GCI becomes more plausible when calculating the output $p_{\mathrm{H_2O}}(x_n)$. It is obvious that at lower operating voltage, $U_{\mathrm{op}} - U_{\mathrm{ampl}}$, the mean current density becomes larger, more fuel gas is consumed, and more steam is

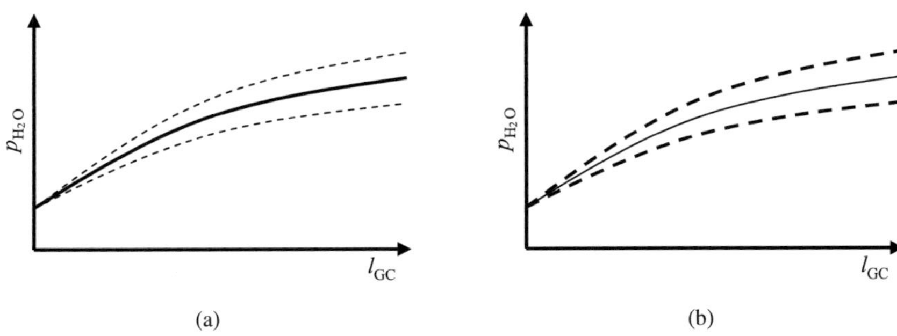

Figure 5.18.: (a) Scheme of the $p_{H_2O}(x_n)$ for high frequencies: it remains unchanged, and (b) for low frequencies equals the maximum displacement, as indicated in figure 5.17.

produced. Then the output $p_{H_2O}(x_N)$ rises. This rise may be small for a small excitation signal and the corresponding change in current density may also be small, but the changes of these quantities show a linear ratio within a certain range, as it is characteristic for any impedance, and that will be specified as GCI, Z_{GCI}.

5.3.7.4. Simple Model for the GCI

In this thesis a very simple approach for the calculation of the GCI is proposed. It neglects gas-phase convection and diffusion along the gas channel and assumes the anode gas to move uniformly from gas inlet to gas outlet without any gradients in the absolute pressure. The model is based on the static macro-model introduced in section 5.3.4. The results of this simple approach will be compared with measurement data in section 5.3.7.5.

Assumptions

Consider a SOFC gas channel in H_2/H_2O operation in an operating point defined by the operation voltage, U_{op}, with an applied current, $I > 0$. With the help of the static macro-model from section 5.3.4, the course of $p_{H_2O}(x_n)$ can be calculated for this operating point as shown in figure 5.17.

High Frequencies

In order to calculate the GCI, it is assumed that for an sinusoidal excitation of the system with a voltage amplitude at a very high frequency, the gas conversion has no influence on the course of $p_{H_2O}(x_n)$. During one positive half-wave, little to no hydrogen is converted to steam and the same amount is reconverted to hydrogen in the subsequent negative half-wave. At high frequencies, these two processes are much faster than the gas transport to the subsequent length element. Hence, the voltage excitation has no influence on the course of $p_{H_2O}(x_n)$, as indicated in figure 5.18(b). This is a very helpful characteristic, because it allows to calculate the GCI at high frequencies with the static macro-model from section 5.3.4. The only difference to the algorithm used there is, that the course of $p_{H_2O}(x_n)$ is known beforehand and only the current density has to be calculated. This is done for two voltage values, $U_{\mathrm{op}} + U_{\mathrm{ampl}}$ and $U_{\mathrm{op}} - U_{\mathrm{ampl}}$. The obtained current values are

$$J_{\mathrm{tot,hi,\infty}} = \sum_{n=1}^{N} J\big(p_{\mathrm{H_2O}}(x_n, U_{\mathrm{op}}), U_{\mathrm{op}} - U_{\mathrm{ampl}}\big) \tag{5.16}$$

and

$$J_{\mathrm{tot,lo,\infty}} = \sum_{n=1}^{N} J\big(p_{\mathrm{H_2O}}(x_n, U_{\mathrm{op}}), U_{\mathrm{op}} + U_{\mathrm{ampl}}\big). \tag{5.17}$$

The amplitude of the current response can be obtained by calculating the difference in the currents obtained by the static macro-model. The impedance is purely ohmic by definition and is calculated via

$$Z_{\mathrm{GCI}}(\omega \to \infty) = \frac{2 \cdot U_{\mathrm{ampl}}}{J_{\mathrm{tot,hi,\infty}} - J_{\mathrm{tot,lo,\infty}}}. \tag{5.18}$$

Low Frequencies

The calculation of the GCI for very low frequencies can be conducted with the static macro-model as well. The impedance converges towards $\mathrm{Im}(Z_{\mathrm{GCI}}) \to 0$, if the frequency is so low that steady state can be assumed, when the sinusoidal excitation reaches its maximum and minimum values $+U_{\mathrm{ampl}}$ and $-U_{\mathrm{ampl}}$, respectively. This way, it is assumed that the time for one excitation period is much larger than the time the gas needs to pass the gas channel. When the voltage reaches its minimum $U_{\mathrm{op}} - U_{\mathrm{ampl}}$, the maximum gas is consumed in the first length element x_1. If the excitation is still at this value after the volume element of gas has passed the whole gas channel and reached the last length element x_N, the whole gas channel can be simulated by the static macro-model with $U_{\mathrm{op}} - U_{\mathrm{ampl}}$, as indicated in figure 5.18(a). The same accounts for the voltage maximum $U_{\mathrm{op}} + U_{\mathrm{ampl}}$. The corresponding expressions are

$$J_{\mathrm{tot,hi,0}} = \sum_{n=1}^{N} J\big(p_{\mathrm{H_2O}}(x_n, U_{\mathrm{op}} - U_{\mathrm{ampl}}), U_{\mathrm{op}} - U_{\mathrm{ampl}}\big) \tag{5.19}$$

and

$$J_{\mathrm{tot,lo,0}} = \sum_{n=1}^{N} J\big(p_{\mathrm{H_2O}}(x_n, U_{\mathrm{op}} + U_{\mathrm{ampl}}), U_{\mathrm{op}} + U_{\mathrm{ampl}}\big). \tag{5.20}$$

Note that the only difference to equations (5.16) and (5.17) is that the course of $p_{\mathrm{H_2O}}(x_n)$ is not taken from the static operating point, U_{op}, but is recalculated for $J_{\mathrm{tot,hi,0}}$ and $J_{\mathrm{tot,lo,0}}$, respectively.

The impedance for very low frequencies is again purely ohmic and is calculated via

$$Z_{\mathrm{GCI}}(\omega \to 0) = \frac{2 \cdot U_{\mathrm{ampl}}}{J_{\mathrm{tot,hi,0}} - J_{\mathrm{tot,lo,0}}}. \tag{5.21}$$

Gas Conversion Impedance

The subtraction of the two resistances yields the contribution of the GCI to the overall resistance of the system:

$$Z_{\mathrm{DC,GCI}} = Z_{\mathrm{GCI}}(\omega \to 0) - Z_{\mathrm{GCI}}(\omega \to \infty). \tag{5.22}$$

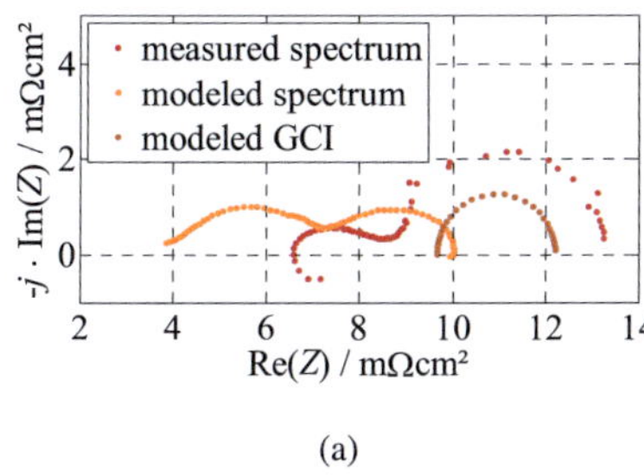

(a)

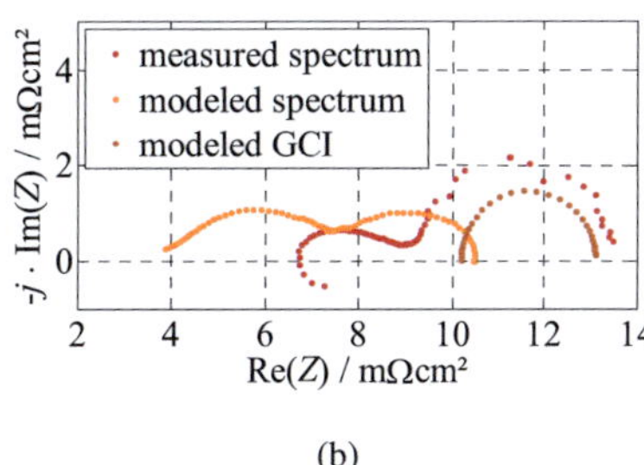

(b)

Figure 5.19.: Comparison of impedance spectra measured on sample S.7b and spectra modeled with the dynamic macro-model parameterized with the spectra measured on sample S.7a plus the modeled GCI: (a) measurement C.1.4 and (b) measurement C.2.2.

As no further information about the dynamics of the GCI, the behavior is assumed to be represented by an RC circuit, the most basic impedance element. At this point, no statement about the characteristic frequency is made.

5.3.7.5. Discussion: Dynamic Macro-Model and Gas Conversion Impedance

It has been stated that the GCI is responsible for the deviation of measured and simulated impedance of the dynamic macro-model. This will be discussed with immediate and further analysis results in this section.

Immediate Results

If the GCI is calculated as explained in section 5.3.7.4 and added to a diagram showing the simulated impedance from section 5.3.6, it fills the gap between measured and simulated impedance with a deviation of $\leq 10\%$ for both $Z_{\mathrm{GCI}}(\omega \rightarrow 0)$ and $Z_{\mathrm{GCI}}(\omega \rightarrow \infty)$, as demonstrated for two examples in figure 5.19.

- **Simulated values $Z_{\mathrm{DC,MM}}$ and $Z_{\mathrm{GCI}}(\omega \rightarrow \infty)$.** These values should be equal as explained in section 5.3.7.4.

- **Simulated value for $Z_{\mathrm{GCI}}(\omega \rightarrow 0)$ and measured value for $Z(\omega \rightarrow 0)$.** These two values should also be equal because the derivation of the GCI in section 5.3.7.4 aims to simulate the Direct Current (DC) resistance of the large-scale system.

This indicates that in fact an additional impedance contribution has to be added for large-scale systems and that the gas conversion in the anode gas channel is the physical process causing this contribution. But due to the poor accuracy, further considerations are necessary.

Further Considerations

Both deviations show the same sign. The modeled DC resistance is smaller than the measured DC resistance. Further, the deviation in figure 5.19(a) is larger than in figure 5.18(b).

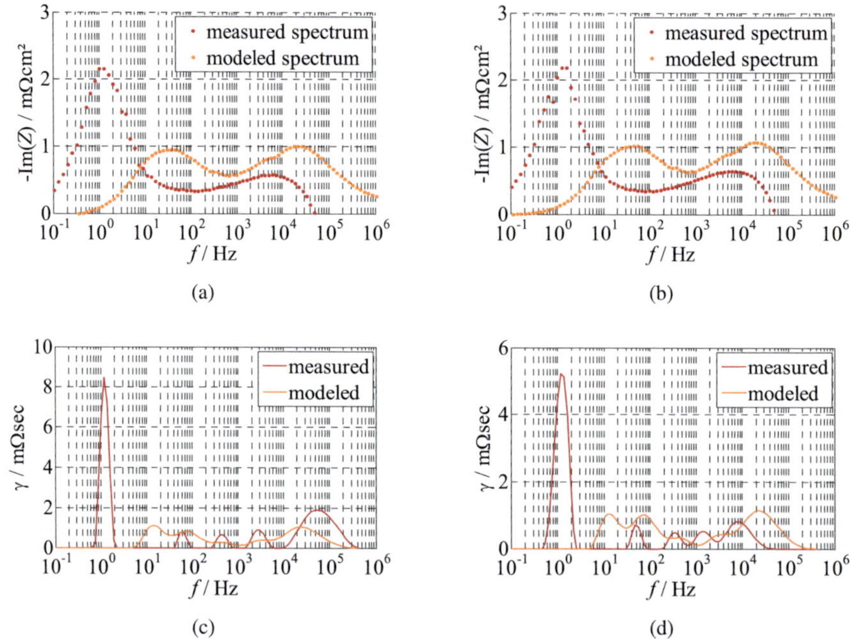

Figure 5.20.: Imaginary parts of the impedance ((a) and (b)) and DRTs ((c) and (d)) for measurement and model (also shown in figure 5.19) of (a) and (c) C.1.4 and (b) and (d) C.2.2.

This is in accordance with the deviations, that can be observed for the simulated C/V curves and their slopes in figure 5.12.

Another fact that has to be mentioned is that the imaginary part of the simulated impedance shows a local maximum in the absolute values around $f = 50\,\mathrm{Hz}$ in figures 5.20(a) and 5.20(b). Near this frequency, at around $f = 100\,\mathrm{Hz}$, the measured impedance shows a local minimum in the absolute imaginary part. By adding another impedance contribution like the proposed Z_{GCI}, this deviation can only get larger and not smaller.

Therefore, this approach is not leading to the desired result and a new gas transport and gas conversion model has to be developed for large-scale SOFC systems.

Further Analysis by DRT and CNLS fit

Figures 5.20(c) and 5.20(d) show DRTs of measurements and simulations, respectively. The high frequent part ($f > 10\,\mathrm{kHz}$) should not be considered, as the measurement shows inductive behavior at high frequencies. At mid-range frequencies ($50\,\mathrm{Hz} < f < 10\,\mathrm{kHz}$), it can be seen that the processes detected in the measurement also appear in the simulated spectra.

At low frequencies ($f < 50\,\mathrm{Hz}$), it can be observed that the gas diffusion process in the anode substrate ($P_{1\mathrm{A}}$, see table 5.1 and [39]) has disappeared in the measurement of the

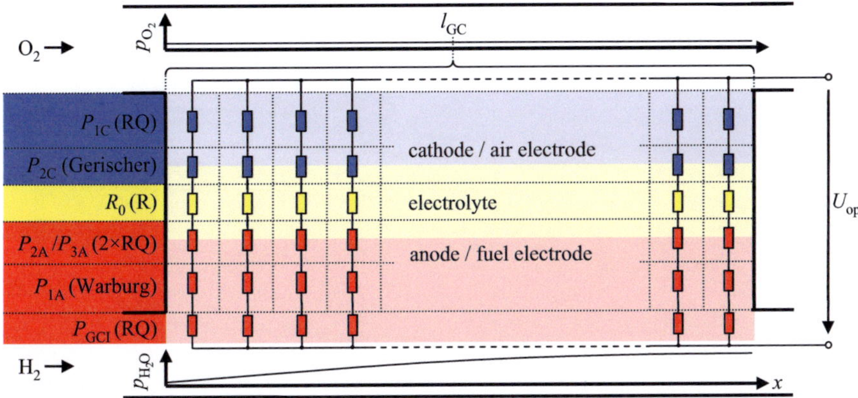

Figure 5.21.: Schematic of the 1D model shown in figure 5.13 but with the GCI as additional process as modeled here.

A_{16} cell ($f \approx 10\,\mathrm{Hz}$). Instead a very narrow process is detected at very low frequencies ($f = 1\ldots3\,\mathrm{Hz}$). This process is not detectable in A_1 cell measurements and will be called P_{GCI} in the following.

A CNLS fit was conducted for all A_{16} impedance measurements in order to identify the characteristics of this process. An RQ circuit could be fitted with very good accuracy to the measurements and the obtained parameters reveal the following facts:

- Very high values for α were obtained for all spectra: $\alpha = 0.97\ldots1.00$. This indicates that the analyzed process shows nearly RC behavior.

- The magnitude of $Z_{\mathrm{DC,GCI}}$ varies between $Z_{\mathrm{DC,GCI}} = 1.5\,\mathrm{m\Omega}$ for $Q_{\mathrm{gas}} = 1000\,\mathrm{ml/min}$, $U_{\mathrm{op}} = 0.85\,\mathrm{V}$, $p_{\mathrm{H_2O}}(0) = 0.2\ldots0.45\,\mathrm{atm}$, and $Z_{\mathrm{DC,GCI}} = 6.8\,\mathrm{m\Omega}$ for $Q_{\mathrm{gas}} = 250\,\mathrm{ml/min}$, $U_{\mathrm{op}} = 1.00\,\mathrm{V}$, $p_{\mathrm{H_2O}}(0) = 0.2\,\mathrm{atm}$.

- The characteristic frequencies clearly indicate a correlation to the anode gas flow rate Q_{gas} and are reciprocally proportional to the time, the gas needs to pass the anodic gas channel for the corresponding flow rate. The results are compiled in table 5.7.

Q_{gas} in ml/min	f_{char} in Hz of Z_{GCI}	τ_{char} in msec	calculated retention time t_{R} in msec
1000	$4.9\ldots5.6$	$28\ldots32$	58
500	$2.3\ldots2.9$	$55\ldots69$	115
250	$1.2\ldots1.3$	$122\ldots132$	230

Table 5.7.: Comparison of gas flow srates (Q_{gas}) and corresponding retention times t_{R} with the characteristic frequencies f_{char} and the time constants τ_{char} obtained by CNLS fit from the spectra measured on sample S.7b.

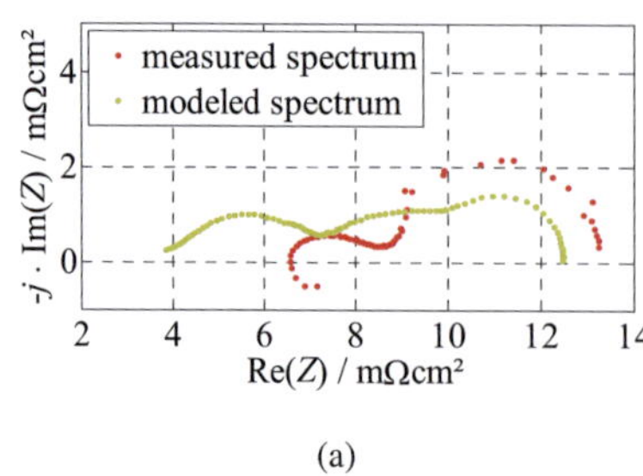

(a)

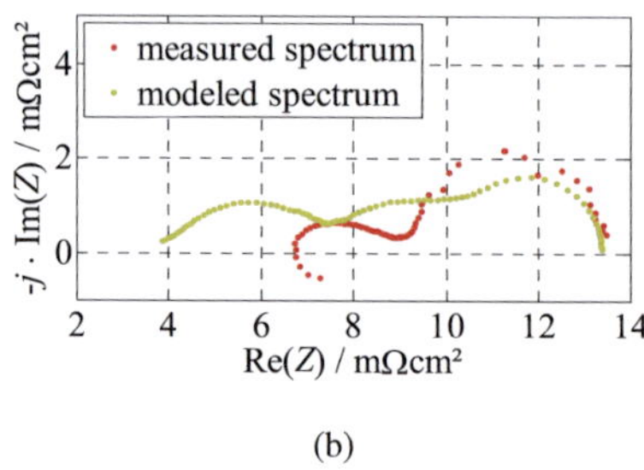

(b)

Figure 5.22.: Measured impedances in comparison with the modeled impedances, for which the GCI was added to the dynamic macro-model derived in section 5.3.6: (a) C.1.4 and (b) C.2.2.

Conclusions: Dynamic Macro-Model and Gas Conversion Impedance

A new process has been identified that only occurs in large-scale SOFC systems, where the p_{H_2O} cannot be assumed to be equal at all positions alongside the gas channel. It has been shown that the time constant of this process is proportional to the time, the gas needs to pass the anodic gas channel. It is therefore attributed to the GCI in the anode gas channel as explained in section 5.3.7.

A formula to calculate the magnitude of the GCI has yet to be derived. Also the coupling with the diffusion in the anode substrate is not fully understood yet. A simple summation of the impedance calculated by the dynamic macro-model and the GCI with the calculated magnitude and the fitted time constant was proposed as dynamic macro-model. The corresponding schematic is shown in figure 5.21.

However, the simulations show that this model does not adequately represent the behavior of the large-scale system, as shown in figure 5.22.

A coupling of GCI and gas diffusion in the anode substrate is suspected to take place in a large-scale SOFC system. Furthermore, it cannot be excluded that the kinetics of the gas flow along the gas channel play a significant role and must be included in the model.

Nevertheless, it could be shown that the dynamic macro-model is a very good basis for the identification of the differences of small-scale and large-scale systems. Further studies should include an extended parameter space for the A_1 and A_{16} cell measurements including lower temperatures, in order to enable a higher quality for the high frequency measurements on the A_{16} cells. At the time of writing, this issue is being analyzed by a diploma thesis at IWE.

6. Summary

This thesis dealt with the electrochemical characterization and modeling of electrochemical energy conversion systems with two prominent systems as examples, the Solid Oxide Fuel Cell (SOFC) and the Lithium-Ion Battery (LIB). Recently, they have gained a lot of attention because of their capability to replace – together with renewable energy sources – fossil and atomic fuels as major sources for the supply of electrical energy in the future.

The focus of this thesis was on the SOFC, a system that is attractive for two reasons. First, it has the potential to turn hydrocarbons into electrical energy with an unchallenged efficiency and can therefore act as an interim solution, until the hydrogen technology and distribution are established. In H_2/H_2O operation, even higher efficiencies are possible. The second reason is its reversibility and the related possibility to act as both fuel cell and electrolyzer cell. This field is relatively new and viable strategies for systems and applications are still being discussed.

In the first part of this thesis, an overview of the general, electrical, chemical and thermal properties of electrochemical systems was given with focus on SOFCs and LIBs.

A fundamental precondition for a successful optimization of such a system is a profound knowledge of the process or processes involved during operation. Measurement and characterization techniques do exist that help to understand the available systems and to establish valuable models. These techniques can be seen as a large toolbox with different measurement techniques to identify the various characteristics of these systems. An introduction to this toolbox of basic measurement techniques that are currently used for the electrochemical characterization was given in the second part of this thesis. This overview was broad but completeness cannot be claimed. Therefore, the field of electrochemical measurements is far too versatile.

Based on these standard techniques for the characterization of SOFCs, amendments have been presented in this thesis that try to fill gaps that are still existing for an impartial characterization of electrochemical systems.

The actual cell temperature during a characterization experiment is one of the operating parameters, that is often not measured adequately. Because of the endothermic reaction in

Solid Oxide Electrolysis Cells (SOECs), the behavior of the cell temperature becomes an even more important issue for the characterization of SOECs. Therefore, a technique was developed in this thesis that does not apply standard temperature sensors but measures the temperature by its intrinsical and unambiguous effects on the conductivity of the electolyte by measuring the impedance in one frequency point (Single Frequency Electrochemical Impedance Spectroscopy (SFEIS)) in section 3.5.3.1. This technique is not completely new and has already been reported for some electrochemical systems, but here it is applied for the first time to a SOFCs.

It was further shown that the dynamics of the evolution of the temperature as a response to a current step can be determined without being obliged to take into account the thermal mass of the sensor, because the sensor is the tested system itself. This technique has been adapted and expanded by a temperature excitation for LIBs by a subsequent work at the Institut für Werkstoffe der Elektrotechnik (IWE) to Electro-Thermal Impedance Spectroscopy (ETIS). In addition, SFEIS was successfully applied to characterize the dynamic change of the impedance of a Lanthanum Strontium Cobaltate (LSC) nano-cathode upon gas variation in section 3.5.3.2.

A complete characterization of a system involves the analysis of the whole frequency range in which a dynamic behavior is detectable. For SOFCs, the slowest processes are still convenient to handle by Electrochemical Impedance Spectroscopy (EIS). This is not the case for LIBs, where very slow solid-state diffusion processes in the sub-millihertz regime are obstructing a fast and complete characterization. Time Domain Measurements (TDMs) in combination with the Fourier transformation are being consulted in order to optimize the measurement time. One elaborate method to reduce the noise and systematic errors (caused by the finite sampling frequency and the finite measurement time) is the TDM derived in section 3.6.2, where an impedance spectrum can be calculated from the system response to an arbitrary excitation signal. Other approaches to calculate the impedance spectrum from time domain data do exist in the literature, however, most of them rely on a pre-defined Equivalent Circuit Model (ECM) to which the time domain data is fitted. This is seen critical because it implies a certain behavior of the system and is not recommended for the characterization of unknown systems, where no ECM can be specified a priori. The technique presented here only takes into account the measured data and the calculated impedance spectrum only represents the observed behavior during measurement. Measurement results of a commercial LIB are shown and both the capability and the limits of this technique are discussed in section 3.6.2.

But even if all relevant parameters are included unambiguously in the measurement data and can be displayed adequately in an impedance spectrum, it is still difficult to extract them with standard identification methods in a consistent way. This becomes even more difficult for the technically relevant systems discussed in this thesis – as compared to simple model systems. However, it was already pointed out in the past that the characterization of a technically relevant system requires the measurement of exactly this complex system, because comparable materials and functional layers may behave differently in another setup. For the non-trivial parameter identification, suitable tools have been developed and are available, but there are cases in which the Complex Nonlinear Least Squares (CNLS) fit fails to converge towards the unique solution in the global minimum of the quality criterion and the Distribution of Relaxation Times (DRT) can only visualize the different processes in a more delicate way. In this case, a combination of those two techniques in an advanced optimization algorithm can yield the desired parameters with both good accuracy and robustness, as shown in section 4.5.4.

Eventually, all the gathered knowledge and the corresponding parameters have to be compiled into an appropriate model. On the basis of a well-proven ECM for small SOFC single cells, a system model was established. It is a distributed parameter model that takes into account the local operating conditions of a small segment within a large-scale system that determine the electrochemical behavior of the corresponding cell segment.

The model yields information about various characteristics of the system. In chapter 5, a number of static and dynamic micro- and macro-models could be derived that can be parameterized by the static operating parameters and the corresponding impedance data. One important result of this chapter is the static behavior model for large-scale systems (macro-model). The accuracy of the simulation results could be confirmed by parameterizing the model with measurement data of a small single cell ($A_1 = 1\,\mathrm{cm}^2$) and by reproducing the measurement results of a large single cell ($A_{16} = 16\,\mathrm{cm}^2$) in section 5.3.4.4.

Furthermore, a detailed electrochemical macro-model was derived in section 5.3.5 that can predict the individual losses of a SOFC system caused by the most important electrochemical processes already identified previously.

Finally, the basis for a dynamic macro-model was laid in order to enable the simulation of large-scale SOFC systems in section 5.3.6. In its current state, its accuracy is not satisfactory and additional measurements and modeling are needed to identify an appropriate dynamic model for a large-scale system in the whole relevant frequency range. Nevertheless, it has already helped to identify the Gas Conversion Impedance (GCI) in both simulation and measurement and will help to further detect critical issues for the up-scaling of electrochemical micro-models. It is thereby not restricted to SOFCs.

On the basis of this modeling approach, a capable diagnosis system can be constructed that has the potential to attribute its current state to desired and undesired operating conditions, to fault-free or faulty operation and to cycle or life-time expectancy.

Even though it has not gained the attention it deserves in this thesis, the Reverse Current Treatment (RCT) has been introduced briefly in section 2.5.2.2. Through a current pulse in negative direction, a well-performing nano-layer between the electrolyte and the anode could be formed in situ on a SOFC single cell. The results of this study have been published and presented several times and have gained a lot of interest while research on this topic is currently going on at IWE and has already produced even more promising results.

These approaches and results together aim at understanding and improving electrochemical energy conversion systems and give impartial numbers for performance, efficiency, and judge their economic plausibility. To answer these questions consistently will be a very important issue for the design of the future supply of electrical energy.

Appendix

A. Measured Samples

	IWE-ID	supplier ID	supplier	remarks
S.1	Z8_86	7905/16	FZJ	ASC with $A = 1\,\mathrm{cm}^2$ active area and $d = 1\,\mathrm{mm}$ for RCT (section 2.5.2.2)
L.1	AA01-0166	SPB333450HS1	Saehan	commercial pouch-cell with a nominal capacity of $Q = 350\,\mathrm{mAh}$ (section 3.4)
S.2	Z8_059	4350	FZJ	ASC with $A = 1\,\mathrm{cm}^2$ active area and $d = 1\,\mathrm{mm}$ for validation of the measurement under load via shunt (section 3.5.1.7)
S.3	Z8_062	07013 121-2	Ceramtec	ASC with $A = 1\,\mathrm{cm}^2$ active area and $d = 1\,\mathrm{mm}$ temperature measurement via SFEIS (section 3.5.3.1)
S.4	Z1_222	10528	FZJ	ASC with $A = 1\,\mathrm{cm}^2$ active area and $d = 1\,\mathrm{mm}$ temperature measurement via SFEIS in SOFC and SOEC mode conducted by J.-C. Njodzefon (section 3.5.3.1)
S.5	Z10_49	PNC-161	IWE	symmetrical LSC cathode $d = 1\,\mathrm{mm}$ on a DKKK CGO pellet conducted by J. Hayd (section 3.5.3.2 and [10])
L.2	MS1	AE452025P4H3R	REVA	commercial pouch cell with nominal capacity of $Q = 120\,\mathrm{mAh}$ conducted by M. Schönleber (section 3.6.2.5)
S.6	Z2_194	12798-14	FZJ	ASC with $A = 1\,\mathrm{cm}^2$ active area and $d = 1\,\mathrm{mm}$, measurement under load for validation of fitting algorithm conducted by A. Kromp (sections 4.2 to 4.5.5)

	IWE-ID	supplier ID	supplier	remarks
S.7a	Z9_131	12384	FZJ	ASC with $A = 1\,cm^2$ active area and $d = 1\,mm$ for micro- and macro model, apart from A identical to S.7b (section 5.3)
S.7b	Z5_166	12417	FZJ	ASC with $A = 16\,cm^2$ active area $d = 1\,mm$ for micro- and macro and model, apart from A identical to S.7a (section 5.3)
S.8	Z9_054	07013 159-1	Ceramtec	ASC with $A = 1\,cm^2$ active area and $d = 1\,mm$ for micro- and macro model (section 5.3)

Table A.1.: List of tested SOFC (S) and LIB (L) samples.

B. Supervised Diploma Theses and Study Projects

- H. Köhler. Analyse des dynamischen Temperaturverhaltens der SOFC unter elektrischer Last. Study project, 2009.

- J. P. Schmidt. Entwicklung eines Diagnose Tools für eine SOFC APU. Diploma thesis, 2009.

- M. Schönleber. Entwicklung eines kombinierten Messverfahrens im Zeit- und Frequenzbereich für Lithium-Ionen-Batterien. Study project, 2010.

- J. Richter. Messung von quasistationären Kennlinien von Batterien und die Simulation der Leerlaufspannung. Diploma thesis, 2011 (co-supervised with J. P. Schmidt).

- M. Schönleber. Development of a Combined State and Parameter Estimation Scheme for Lithium-Ion Batteries. Diploma thesis (external), 2011.

C. Publications

- D. Klotz, A. Weber and E. Ivers-Tiffée, 'Thermal Behaviour of SOFC Single Cells in Operation with Dynamic Loads', in R. Steinberger-Wilckens and U. Bossel (Eds.), Proceedings of the 8th European Solid Oxide Fuel Cell Forum, B0909 (2008).

- D. Klotz, A. Leonide and E. Ivers-Tiffée, 'Recovery of Anode Performance by Reverse Current Treatment', ECS Transactions 25, pp. 2049-2056 (2009).

- D. Klotz, A. Weber and E. Ivers-Tiffée, 'Dynamic Electrochemical Model for SOFC Stacks', ECS Transactions 25, pp. 1331-1340 (2009).

- D. Klotz, J. P. Schmidt, A. Weber and E. Ivers-Tiffée, 'Dynamic Electrochemical Model for SOFC Stacks', in First International Conference on Materials for Energy, Extended Abstracts - Book A, Frankfurt am Main: DECHEMA, pp. 102-104 (2010).

- D. Klotz, A. Leonide and E. Ivers-Tiffée, 'Increase of Anode Performance of SOFC by Reverse Current Treatment', in First International Conference on Materials for Energy, Extended Abstracts - Book A, Frankfurt am Main: DECHEMA e. V., pp. 75-77 (2010).

- T. Chrobak, M. Ender, J. Illig, J. P. Schmidt, D. Klotz and E. Ivers-Tiffée, 'Studies on $LiFePO_4$ as Cathode Materials in Li-Ion-Batteries', in First International Conference on Materials for Energy, Extended Abstracts - Book B: DECHEMA, pp. 593- 595 (2010).

- D. Klotz, J. P. Schmidt, A. Weber and E. Ivers-Tiffée, 'Dynamic Electrochemical Model for SOFC-Stacks', in J. T. S. Irvine and U. Bossel (Eds.), Proceedings of the 9th European Solid Oxide Fuel Cell Forum, pp. 1-6 (2010).

- T. Yamamoto, H. Morita, Y. Mugikura, D. Klotz, A. Leonide, A. Weber and E. Ivers-Tiffée, 'Investigation of SOFC Performance and Durability Evaluations', in J. T. S. Irvine and U. Bossel (Eds.), Proceedings of the 9th European Solid Oxide Fuel Cell Forum, pp. 7-86 - 7-95 (2010).

- D. Klotz, B. Butz, A. Leonide, D. Gerthsen and E. Ivers-Tiffée, 'Increase of Anode Performance of SOFC by Reverse Current Treatment', ECS Transactions 28, pp. 141-150 (2010).

- J. Illig, T. Chrobak, M. Ender, J. P. Schmidt, D. Klotz and E. Ivers-Tiffée, 'Studies on $LiFePO_4$ as Cathode Material in Li-Ion Batteries', ECS Transactions 28, pp. 3-17 (2010).

- J. P. Schmidt, T. Chrobak, M. Ender, J. Illig, D. Klotz and E. Ivers-Tiffée, 'Studies on $LiFePO_4$ as Cathode Material using Impedance Spectroscopy', Journal of Power Sources 196, pp. 5342-5348 (2010).

- J. Illig, T. Chrobak, D. Klotz and E. Ivers-Tiffée, 'Evaluation of the Rate Determining Processes for $LiFePO_4$ as Cathode Material in Lithium-Ion-Batteries', ECS Transactions 33, pp. 3-15 (2011).

- D. Klotz, B. Butz, A. Leonide, J. Hayd, D. Gerthsen and E. Ivers-Tiffée, 'Performance Enhancement of SOFC Anode Through Electrochemically Induced Ni/YSZ Nanostructures', Journal of the Electrochemical Society 158, pp. B587-B595 (2011).

- E. Ivers-Tiffée, J. Hayd, D. Klotz, A. Leonide, F. Han and A. Weber, 'Performance Analysis and Development Strategies for Solid Oxide Fuel Cells', ECS Transactions 35, pp. 1965-1973 (2011).

- J. P. Schmidt, D. Manka, D. Klotz and E. Ivers-Tiffée, 'Investigation of the Thermal Properties of a Li-Ion Pouch-Cell by Electrothermal Impedance Spectroscopy', Journal of Power Sources 196, pp. 8140-8146 (2011).

- D. Klotz, M. Schönleber, J. P. Schmidt and E. Ivers-Tiffée, 'New Approach for the Calculation of Impedance Spectra out of Time Domain Data', Electrochimica Acta 56, pp. 8763-8769 (2011).

- J. P. Schmidt, D. Klotz, J. Richter and E. Ivers-Tiffée, 'Insertion Potentials of Li-Ion Battery Electrodes quickly Identified by a Mathematical Transformation of OCV Curves', in Dominique Guyomard (Ed.), Proceedings of the Lithium Batteries Discussion 2011, p. P12 (2011).

- J. P. Schmidt, J. Richter, D. Klotz and E. Ivers-Tiffée, 'Beneficial Use of a Virtual Reference Electrode for the Determination of SOC Dependent Half Cell Potentials', ECS Transactions 41, pp. 1-8 (2012).

- D. Klotz, J. P. Schmidt, A. Kromp, A. Weber and E. Ivers-Tiffée, 'The Distribution of Relaxation Times as Beneficial Tool for Equivalent Circuit Modeling of Fuel Cells and Batteries', ECS Transactions 41, pp. 25-33 (2012).

- J.-C. Njodzefon, D. Klotz, A. Leonide, A. Kromp, A. Weber and E. Ivers-Tiffée, 'Electrochemical Studies on Anode Supported Solid Oxide Electrolyzer Cells', ECS Transactions 41, pp. 113-122 (2012).

- D. Klotz, J. Szasz, A. Weber and E. Ivers-Tiffée, 'Nano-Structuring of SOFC Anodes by Reverse Current Treatment', ECS Transactions 45, pp. 241-249 (2012).

- D. Klotz, J.-C. Njodzefon, A. Weber and E. Ivers-Tiffée, 'Current-Voltage and Temperature Characteristics of Anode Supported Solid Oxide Electrolyzer Cells (SOEC)', ECS Transactions 45, pp. 523-530 (2012).

- J. Illig, M. Ender, T. Chrobak, J. P. Schmidt, D. Klotz and E. Ivers-Tiffée, 'Separation of Charge Transfer and Contact Resistance in $LiFePO_4$-Cathodes by Impedance Modeling', Journal of the Electrochemical Society 159, pp. A952-A960 (2012).

D. Conference Contributions

- D. Klotz, A. Weber, E. Ivers-Tiffée, Poster presentation by D. Klotz: 'SOFC Single Cells Operated under Dynamic Loads', 11th UECT Ulm ElectroChemical Talks 2008 (Ulm, Germany), 10.06. - 12.06.2008.

- D. Klotz, A. Weber, E. Ivers-Tiffée, Poster presentation by D. Klotz: 'Thermal Behaviour of SOFC Single Cells in Operation with Dynamic Loads', 8th EUROPEAN SOFC FORUM (Lucerne, Switzerland), 30.06. - 04.07.2008.

- D. Klotz, A. Weber, B. Rüger, E. Ivers-Tiffée, Oral presentation by D. Klotz: 'Betrieb von Hochtemperatur-Brennstoffzellen (SOFC) unter transienten Lastbedingungen', f-cell 2008 (Stuttgart, Germany), 29.09. - 30.09.2008.

- A. Weber, H. Timmermann, D. Klotz, M.J. Heneka, E. Ivers-Tiffée, Oral presentation by A. Weber: 'Cycling Effects', International Workshop on Degradation Issues in Fuel Cells (ZSW Ulm, Germany), 06.10. - 07.10.2008.

- A. Weber, D. Klotz, V. Sonn, E. Ivers-Tiffée, Oral presentation by A. Weber: 'Impedance Spectroscopy as a Diagnosis Tool for SOFC Stacks and Systems', International Symposium on Diagnostics Tools for Fuel Cell Technologies (Trondheim, Norway), 23.06. - 24.06.2009.

- D. Klotz, A. Weber, E. Ivers-Tiffée, Oral presentation by D. Klotz: 'EIS Analysis of YSZ-Electrolyte Temperature Distribution in Operation with Dynamic Loads, 17th International Conference on Solid State Ionics (SSI-17) (Toronto, Canada), 28.06. - 03.07.2009.

- D. Klotz, J.P. Schmidt, A. Weber, E. Ivers-Tiffée, Poster presentation by D. Klotz: 'Dynamic Electrochemical Model for SOFC Stacks, SOFC XI (Vienna, Austria), 04.10. - 09.10.2009.

- D. Klotz, A. Leonide, J. Hayd, E. Ivers-Tiffée, Poster presentation by D. Klotz: 'Recovery of Anode Performance by Reverse Current Treatment', SOFC XI (Vienna, Austria), 04.10. - 09.10.2009.

- D. Klotz, A. Leonide, E. Ivers-Tiffée, Oral presentation by D. Klotz: 'Increase of Anode Performance of SOFC by Reverse Current Treatment', 217th Meeting of The Electrochemical Society (Vancouver, Canada), 25.04. - 30.04.2010.

- J. Illig, T. Chrobak, M. Ender, J.P. Schmidt, D. Klotz, E. Ivers-Tiffée, Oral presentation by J. Illig: 'Studies on $LiFePO_4$ as Cathode Material in Li-Ion Batteries', 217th Meeting of The Electrochemical Society (Vancouver, Canada), 26.04. - 30.04.2010.

- J.P. Schmidt, T. Chrobak, M. Ender, J. Illig, D. Klotz, E. Ivers-Tiffée, Poster presentation by J. P. Schmidt: 'Studies on $LiFePO_4$ as Cathode Material using Impedance Spectroscopy', 12th Ulm ElectroChemical Talks (UECT) (Ulm, Germany), 16.06. - 17.06.2010.

- M. Ender, T. Chrobak, J. Illig, J.P. Schmidt, D. Klotz, E. Ivers-Tiffée, Poster presentation by M. Ender: 'Identification of Reaction Mechanisms in Lithium-Ion Cells by Deconvolution of Electrochemical Impedance Spectra', 15th International Meeting on Lithium Batteries (Montréal, Canada), 27.06. - 02.07.2010.

- D. Klotz, J.P. Schmidt, A. Weber, E. Ivers-Tiffée, Poster presentation by D. Klotz: 'Dynamic Electrochemical Model for SOFC Stacks', 9th EUROPEAN SOFC FORUM (Lucerne, Switzerland), 29.06. - 02.07.2010.

- T. Yamamoto, H. Morita, Y. Mugikura, D. Klotz, A. Leonide, A. Weber, E. Ivers-Tiffée, Poster presentation by T. Yamamoto: 'Investigation of SOFC Performance and Durability Evaluations', 9th EUROPEAN SOFC FORUM (Lucerne, Switzerland), 29.06. - 02.07.2010.

- D. Klotz, J.P. Schmidt, A. Weber, E. Ivers-Tiffée, Poster presentation by D. Klotz: 'Dynamic Electrochemical Model for SOFC Stacks', First International Conference on Materials for Energy (Karlsruhe, Germany), 04.07. - 08.07.2010.

- D. Klotz, A. Leonide, E. Ivers-Tiffée, Oral presentation by D. Klotz: 'Increase of Anode Performance of SOFC by Reverse Current Treatment', First International Conference on Materials for Energy (Karlsruhe, Germany), 04.07. - 08.07.2010.

- T. Chrobak, M. Ender, J. Illig, J.P. Schmidt, D. Klotz, E. Ivers-Tiffée, Oral presentation by T. Chrobak: 'Studies on $LiFePO_4$ as Cathode Material in Li-Ion-Batteries', First International Conference on Materials for Energy (Karlsruhe, Germany), 05.07. - 08.07.2010.

- J.P. Schmidt, D. Klotz, A. Weber, E. Ivers-Tiffée, Poster presentation by J. P. Schmidt: 'Dynamic Electrochemical Model for SOFC Stacks', 61st Annual ISE Meeting (Nice, France), 26.09. - 01.10.2010.

- J. Illig, T. Chrobak, D. Klotz, E. Ivers-Tiffée, Oral presentation by J. Illig: 'Evaluation of the Rate Determining Processes for $LiFePO_4$ as Cathode Material in Lithium-Ion-Batteries', 218th Meeting of The Electrochemical Society (Las Vegas, USA), 10.10. - 15.10.2010.

- E. Ivers-Tiffee, J. Hayd, D. Klotz, A. Leonide, F. Han and A. Weber, Oral presentation by E. Ivers-Tiffée: 'Performance Analysis and Development Strategies for Solid Oxide Fuel Cells', 219th Meeting of The Electrochemical Society (Montreal, Canada), 01.05. - 06.05.2011.

- J. Illig, M. Ender, T. Chrobak, J.P. Schmidt, D. Klotz, E. Ivers-Tiffée, Oral presentation by J. Illig: 'Evaluation of $LiFePO_4$-Cathodes by Impedance Modeling', Lithium Batteries Discussion 2011 (Arcachon, France), 12.06. - 17.06.2011.

- J. P. Schmidt, D. Klotz, J. Richter, E. Ivers-Tiffée, Oral presentation by J. P. Schmidt: 'Insertion Potentials of Li-Ion Battery Electrodes quickly Identified by a Mathematical Transformation of OCV Curves', Lithium Batteries Discussion 2011 (Arcachon, France), 12.06. - 17.06.2011.

- J. P. Schmidt, D. Klotz, J. Richter, E. Ivers-Tiffée, Poster presentation by D. Klotz: 'Fast Identification of the Insertion Potentials of Li-Ion Battery Electrodes by a Mathematical Transformation of OCV-Curves', 62nd Annual ISE Meeting (Niigata, Japan), 11.09. - 16.09.2011.

- D. Klotz, M. Schönleber, J. P. Schmidt, E. Ivers-Tiffée, Poster presentation by M. Schönleber: 'New Approach for the Calculation of Impedance Spectra out of Time Domain Data', 62nd Annual ISE Meeting (Niigata, Japan), 11.09. - 16.09.2011.

- D. Klotz, J. P. Schmidt, A. Kromp, A. Weber and E. Ivers-Tiffée, Oral presentation by D. Klotz: 'The Distribution of Relaxation Times as Beneficial Tool for Equivalent Circuit Modeling of Fuel Cells and Batteries', 220th Meeting of The Electrochemical Society (Boston, USA), 09.10. - 14.10.2011.

- J.-C. Njodzefon, D. Klotz, A. Leonide, A. Kromp, A. Weber and E. Ivers-Tiffée, Oral presentation by D. Klotz: 'Electrochemical Studies on Anode Supported Solid Oxide Electrolyzer Cells', 220th Meeting of The Electrochemical Society (Boston, USA), 09.10. - 14.10.2011.

- J.P. Schmidt, J. Richter, D. Klotz, E. Ivers-Tiffée, Oral presentation by J. P. Schmidt: 'A Virtual Reference Electrode Suitable for the Non-Destructive Determination of SOC Dependent Half Cell Potentials', 220th Meeting of The Electrochemical Society (Boston, USA), 09.10. - 14.10.2011.

- D. Klotz, J. P. Schmidt, M. Schönleber, E. Ivers-Tiffée, Oral presentation by D. Klotz: 'Derivation of Impedance Spectra out of Time Domain Data', 221th Meeting of The Electrochemical Society (Seattle, USA), 06.05. - 11.05.2012.

- D. Klotz, J. Szasz, A. Weber and E. Ivers-Tiffée, Oral presentation by D. Klotz: 'Nano-Structuring of SOFC Anodes by Reverse Current Treatment', 221th Meeting of The Electrochemical Society (Seattle, USA), 06.05. - 11.05.2012.

- D. Klotz, J.-C. Njodzefon, A. Weber and E. Ivers-Tiffée, Oral presentation by D. Klotz: 'Current-Voltage and Temperature Characteristics of Anode Supported Solid Oxide Electrolyzer Cells (SOEC)', 221th Meeting of The Electrochemical Society (Seattle, USA), 06.05. - 11.05.2012.

- M. Schönleber, J. P. Schmidt, D. Klotz, A. Weber and E. Ivers-Tiffée, Poster presentation by M. Schönleber: 'Applicability and Restrictions for Time-Domain Methods yielding Impedance Spectra', 16th International Meeting on Lithium Batteries (Jeju, Korea), 17.06. - 22.06.2012.

- J. P. Schmidt, M. Schönleber, D. Klotz, J. Illig, A. Weber and E. Ivers-Tiffée, Poster presentation by J. P. Schmidt: 'Investigation of the μHz-Range in Impedance Spectra of Li-Ion Cells, 16th International Meeting on Lithium Batteries (Jeju, Korea), 17.06. - 22.06.2012.

E. Symbols

Symbol	Description	Unit/Value
A	area	cm^2
ASR	area specific resistance	Ωcm^2
C	capacity	F
E	energy	J
$F\{signal\}$	Fourier transformation of signal	-
ΔG	change in Gibbs free energy	J/mol
ΔH	change in enthalpy	J/mol
I	static current	A
J	static current density	A/cm^2
L	inductivity	H
M	molar mass	$\frac{\text{kg}}{\text{mol}}$
N_A	Avogadro constant	$6.022 \cdot 10^{23}\,\frac{1}{\text{mol}}$
P	power (density)	W or W/cm^2
Q	charge	Ah
Q_gas	gas flow rate	ml/min
R	ideal gas constant	$8.314\,\frac{\text{J}}{\text{Kmol}}$
R_i	resistance of component i	Ω
ΔS	change in entropy	J/Kmol
T	temperature	K
U	voltage	V
V	volume	ml
V_m	molar volume	$22.4\,\frac{1}{\text{mol}}$

Symbol	Description	Unit/Value
Y	admittance	$\frac{1}{\Omega}$
Z	impedance	Ω or Ωcm^2
$\text{Re}\,(Z)$	real part of the impedance	Ω or Ωcm^2
$\text{Im}\,(Z)$	imaginary part of the impedance	Ω or Ωcm^2
Z_{DC}	DC resistance	Ω or Ωcm^2
a	chemical activity	-
$a(t)$	arbitrary signal	-
d	thickness	m
e	elementary charge	$1.602 \cdot 10^{-19}$ C
f	frequency	Hz
$i(t)$	dynamic current	A
j	imaginary unit	$\sqrt{-1}$
k	Boltzmann constant	$1.381 \cdot 10^{-23}\,\frac{\text{J}}{\text{K}}$
l	length	m
m	mass	kg
p_0	ambient pressure	1 atm
p_n	partial pressure of gas component n	atm
t	time	sec
$u(t)$	dynamic voltage	V
α	exponent	-
δ	stochiometric factor	-
η	efficiency	-
η_i	overpotential of process i	V
γ	resistance load per frequency	Ωsec or $\Omega seccm^2$
μ	chemical potential	V
$\bar{\mu}$	electrochemical potential	V
ω	angular frequency	$\frac{\text{rad}}{\text{sec}}$
σ	conductivity	$\frac{1}{\Omega m}$
τ	characteristic time constant	sec
θ	phase angle of a complex number	-
φ	electrical potential	V

F. Acronyms

AC	Alternating Current
AD	Analag Digital
ADIS	Analyzing the Difference in Impedance Spectra
AFL	Anode Functional Layer
ASC	Anode Supported Cell
ASR	Area Specific Resistance
BMS	Battery Management System
C/V	Current/Voltage
CA	Cyclic Amperometry

CCCV	Constant Current Constant Voltage
CCM	Constant Current Measurement
CFD	Computational Fluid Dynamics
CHP	Combined Heat and Power
CI	Current Interrupt
CNLS	Complex Nonlinear Least Squares
CPE	Constant Phase Element
CRIEPI	Central Research Institute of Electric Power Industry
CSTR	Continuously Stirred Tank Reactor
CV	Cyclic Voltammetry
DA	Digital Analag
DC	Direct Current
DEIS	Dynamic Electrochemical Impedance Spectroscopy
DFT	Discrete Fourier Transformation
DIA	Differential Impedance Analysis
DMFC	Direct Methanol Fuel Cell
DRT	Distribution of Relaxation Times
DVA	Differential Voltage Analysis
ECM	Equivalent Circuit Model
ECU	Electronic Control Unit
EIS	Electrochemical Impedance Spectroscopy
ETIS	Electro-Thermal Impedance Spectroscopy
FEM	Finite Elements Method
FFT	Fast Fourier Transformation
FIB	Focused Ion Beam
FRA	Frequency Response Analyzer
FT-EIS	Fourier Transformation Electrochemical Impedance Spectroscopy
FZJ	Forschungszentrum Jülich
GCI	Gas Conversion Impedance
GDC	Gadolinium Doped Ceria
GITT	Galvanostatic Intermittent Titration Technique
GPIB	General Purpose Interface Bus
ICA	Incremental Capacity Analysis
IRS	Institut für Regelungs- und Steuerungssysteme
IWE	Institut für Werkstoffe der Elektrotechnik
JSPS	Japanese Society for the Promotion of Science
KIT	Karlsruhe Institute of Technology
LIB	Lithium-Ion Battery

LSC   Lanthanum Strontium Cobaltate
LSCF   Lanthanum Strontium Cobalt Ferrite
LSM   Lanthanum Strontium Manganite
LSV   Linear Sweep Voltammetry
LZO   Lanthanum Zirconia

MIEC   Mixed Ionic Electronic Conducting

NLEIS   Nonlinear Electrochemical Impedance Spectroscopy
NMC   Nickel Manganese Cobaltite
NSI   Nano-Structured Interlayer
NTC   Negative Temperature Coefficient

OCV   Open Circuit Voltage

PEM   Proton Exchange Membrane
PITT   Potential Intermittent Titration Technique
PPD   (equally spaced) Points Per Decade
PTC   Positive Temperature Coefficient

RCT   Reverse Current Treatment
RVE   Representative Volume Element

SEM   Scanning Electron Microscopy
SFEIS   Single Frequency Electrochemical Impedance Spectroscopy
SOC   State of Charge
SOEC   Solid Oxide Electrolysis Cell
SOFC   Solid Oxide Fuel Cell
SOH   State of Health

TDM   Time Domain Measurement
TEM   Transmission Electron Microscopy
TLM   Transmission Line Model
TPB   Three Phase Boundary

USB   Universal Serial Bus

YSZ   Yttria Stabilized Zirconia

Bibliography

[1] A. J. Bard and L. R. Faulkner. *Electrochemical Methods: Fundamentals and Applications*. Wiley, New York, 2. edition, 2001.

[2] J. S. Newman and K. E. Thomas-Alyea. *Electrochemical Systems*. Wiley, Hoboken, NJ [u.a.], 3. edition, 2004.

[3] C. H. Hamann, A. Hamnett, and W. Vielstich. *Electrochemistry*. Wiley-VCH, Weinheim [u.a.], 2. edition, 2007.

[4] P. W. Atkins and J. De Paula. *Physical Chemistry*. Freeman, New York, 7. edition, 2002.

[5] A. Hamnett. The Components of an Electrochemical Cell. In *Handbook of Fuel Cells*, chapter 1. John Wiley & Sons, Ltd, 2003.

[6] D. Linden. Basic Concepts. In *Handbook of Batteries*, chapter 1. McGraw-Hill, 3. ed. edition, 2002.

[7] R. A. Huggins. *Advanced Batteries : Materials Science Aspects*. Springer, New York, NY, 2009.

[8] D. Linden and Th. B. Reddy, editors. *Lindens Handbook of Batteries*. McGraw-Hill, New York, NY [u.a.], 4. ed. edition, 2011.

[9] L. Carrette, K. A. Friedrich, and U. Stimming. Fuel Cells - Fundamentals and Applications. *Fuel Cells*, 1(1):5–39, 2001.

[10] J. Hayd. *Nanoskalige Kathoden für den Einsatz in Festelektrolyt-Brennstoffzellen bei abgesenkten Betriebstemperaturen*. PhD thesis, Karlsruher Institut für Technologie (KIT), 2012.

[11] R. Steinberger-Wilckens, L. Blum, H.-P. Buchkremer, S. Gross, B. De Haart, K. Hilpert, H. Nabielek, W. Quadakkers, U. Reisgen, R. W. Steinbrech, and F. Tietz. Overview of the Development of Solid Oxide Fuel Cells at Forschungszentrum Juelich. *International Journal of Applied Ceramic Technology*, 3(6):470–476, 2006.

[12] G. M. Ehrlich. Lithium-Ion Batteries. In *Handbook of Batteries*, chapter 35. McGraw-Hill, 3. edition, 2002.

[13] J.-C. Njodzefon, D. Klotz, A. Leonide, A. Kromp, A. Weber, and E. Ivers-Tiffée. Electrochemical Studies on Anode Supported Solid Oxide Electrolyzer Cells. *ECS Transactions*, 41(33):113–122, 2012.

[14] E. Ivers-Tiffée. Brennstoffzellen und Batterien. Lecture notes, Institut für Werkstoffe der Elektrotechnik (IWE), Karlsruher Institut für Technologie (KIT), 2009.

[15] N. N. Greenwood and A. Earnshaw. *Chemistry of the Elements.* Elsevier Butterworth-Heinemann, Amsterdam, 2. edition, 2005.

[16] W. Vielstich. Ideal and Effective Efficiencies of Cell Reactions and Comparison to Carnot Cycles. In *Handbook of Fuel Cells*, chapter 4. John Wiley & Sons, Ltd, 2010.

[17] J. O. Besenhard and M. Winter. Insertion Reactions in Advanced Electrochemical Energy Storage. *Pure and Applied Chemistry*, 70(3):603–608, 1998.

[18] A. Leonide. *SOFC Modelling and Parameter Identification by Means of Impedance Spectroscopy.* PhD thesis, Karlsruher Institut für Technologie (KIT), Karlsruhe, 2010.

[19] D. Linden. Factors Affecting Battery Performance. In *Handbook of Batteries*, chapter 3. McGraw-Hill, 3. ed. edition, 2002.

[20] M. A. Roscher, J. Vetter, and D. U. Sauer. Characterisation of Charge and Discharge Behaviour of Lithium Ion Batteries with Olivine based Cathode Active Material. *Journal of Power Sources*, 191(2):582 – 590, 2009.

[21] E. Barsoukov and J. R. Macdonald. *Impedance Spectroscopy.* John Wiley & Sons, Inc., 2005.

[22] M. E. Orazem nad B. Tribollet. *Electrochemical Impedance Spectroscopy.* The Electrochemical Society series. Wiley, Hoboken, NJ, 2008.

[23] A. Lasia. Electrochemical Impedance Spectroscopy and its Applications. In *Modern Aspects of Electrochemistry*, volume 32, pages 143–248. Ieee, 1999.

[24] C. Gerthsen, H. O. Kneser, and H. Vogel. *Physik : Ein Lehrbuch zum Gebrauch neben Vorlesungen.* Springer, Berlin [u.a.], 15. edition, 1986.

[25] U. Tietze and Ch. Schenk. *Halbleiter-Schaltungstechnik.* Springer, Berlin, 13. edition, 2010.

[26] O. Föllinger, F. Dörrscheidt, and M. Klittich, editors. *Regelungstechnik : Einführung in die Methoden und ihre Anwendung.* Hüthig, Heidelberg, 8. edition, 1994.

[27] J. P. Schmidt, T. Chrobak, M. Ender, J. Illig, D. Klotz, and E. Ivers-Tiffée. Studies on LiFePO$_4$ as Cathode Material using Impedance Spectroscopy. *Journal of Power Sources*, 196(12):5342 – 5348, 2011.

[28] J. P. Schmidt, J. Richter, D. Klotz, and E. Ivers-Tiffée. Beneficial Use of a Virtual Reference Electrode for the Determination of SOC Dependent Half Cell Potentials. *ECS Transactions*, 41(14):1–8, 2012.

[29] M. J. Heneka. *Alterung der Festelektrolyt-Brennstoffzelle unter thermischen und elektrischen Lastwechseln.* PhD thesis, Universität Karlsruhe, 2006.

[30] M. J. Heneka and E. Ivers-Tiffee. Degradation of SOFC Single Cells under severe Current Cycles. *Solid State Electrochemistry. Proceedings. 26. Risø International Symposium on Materials Science*, page 215, 2005.

[31] C. Peters. *Grain-Size Effects in Nanoscaled Electrolyte and Cathode Thin Films for Solid Oxide Fuel Cells (SOFC)*. PhD thesis, Karlsruher Institut für Technologie (KIT), Karlsruhe, 2009.

[32] B. Scherrer. *Composition, Microstructures and Electrical Properties of Yttria-stabilized-Zirconia Thin films made by Spray Pyrolysis*. PhD thesis, ETH Zurich, 2012.

[33] D. Johnson. *Zview2 Help*, 2002.

[34] M. R. S. Abouzari, F. Berkemeier, G. Schmitz, and D. Wilmer. On the Physical Interpretation of Constant Phase Elements. *Solid State Ionics*, 180(14):922 – 927, 2009.

[35] U. Moissl, P. Wabel, St. Leonhardt, and R. Isermann. Modellbasierte Analyse von Bioimpedanz-Verfahren. *Automatisierungstechnik, at*, 52:270–279, Januar 2004.

[36] H. Schichlein. *Experimentelle Modellbildung für die Hochtmperatur-Brennstoffzelle SOFC*. PhD thesis, Universität Karlsruhe, 2003.

[37] M. S. Haschka. *Online-Identifikation fraktionaler Impedanzmodelle für die Hochtemperaturbrennstoffzelle SOFC*. PhD thesis, Universität Karlsruhe, 2008.

[38] S. Buller. *Impedance-based Simulation Models for Energy Storage Devices in Advanced Automotive Power Systems*. PhD thesis, RWTH Aachen, 2003.

[39] A. Leonide, V. Sonn, A. Weber, and E. Ivers-Tiffée. Evaluation and Modeling of the Cell Resistance in Anode-Supported Solid Oxide Fuel Cells. *Journal of The Electrochemical Society*, 155(1):B36–B41, 2008.

[40] J. Bisquert, G. Garcia-Belmonte, F. Fabregat-Santiago, and A. Compte. Anomalous Transport Effects in the Impedance of Porous Film Electrodes. *Electrochemistry Communications*, 1(9):429 – 435, 1999.

[41] J. Bisquert. Influence of the Boundaries in the Impedance of Porous Film Electrodes. *Phys. Chem. Chem. Phys.*, 2:4185–4192, 2000.

[42] B. A. Boukamp and H. J. M. Bouwmeester. Interpretation of the Gerischer Impedance in Solid State Ionics. *Solid State Ionics*, 157(1-4):29 – 33, 2003.

[43] S. B. Adler. Mechanism and Kinetics of Oxygen Reduction on Porous $La_{1-x}Sr_xCoO_{3-d}$ Electrodes. *Solid State Ionics*, 111(1-2):125 – 134, 1998.

[44] J. Love, S. Amarasinghe, D. Selvey, X. Zheng, and L. Christiansen. Development of SOFC Stacks at Ceramic Fuel Cells Limited. *ECS Transactions*, 25(2):115–124, 2009.

[45] M. A. Roscher and D. U. Sauer. Dynamic Electric Behavior and Open-Circuit-Voltage Modeling of $LiFePO_4$-based Lithium Ion Secondary Batteries. *Journal of Power Sources*, 196(1):331 – 336, 2011.

[46] Y. Zhu and Ch. Wang. Strain Accommodation and Potential Hysteresis of LiFePO$_4$ Cathodes during Lithium Ion Insertion/Extraction. *Journal of Power Sources*, 196(3):1442 – 1448, 2011.

[47] K. Xu. Nonaqueous Liquid Electrolytes for Lithium-Based Rechargeable Batteries. *ChemInform*, 35(50):4303–4417, 2004.

[48] H. Timmermann. *Untersuchungen zum Einsatz von Reformat aus flüssigen Kohlenwasserstoffen in der Hochtemperaturbrennstoffzelle SOFC*. PhD thesis, Karlsruher Institut für Technologie (KIT), 2010.

[49] Y. Guo. SAFETY | Thermal Runaway. In Editor in Chief: Jürgen Garche, editor, *Encyclopedia of Electrochemical Power Sources*, pages 241 – 253. Elsevier, Amsterdam, 2009.

[50] J. P. Schmidt, D. Manka, D. Klotz, and E. Ivers-Tiffée. Investigation of the Thermal Properties of a Li-Ion Pouch-Cell by Electrothermal Impedance Spectroscopy. *Journal of Power Sources*, 196(19):8140 – 8146, 2011.

[51] E. Barsoukov, J. H. Jang, and H. Lee. Thermal Impedance Spectroscopy for Li-Ion Batteries using Heat-Pulse Response Analysis. *Journal of Power Sources*, 109(2):313 – 320, 2002.

[52] H. S. Spacil. 3 503 809, 1970. US patent.

[53] A. Ioselevich, A. A. Kornyshev, and W. Lehnert. Degradation of Solid Oxide Fuel Cell Anodes Due to Sintering of Metal Particles. *Journal of The Electrochemical Society*, 144(9):3010–3019, 1997.

[54] A. Ioselevich, A. A. Kornyshev, and W. Lehnert. Statistical Geometry of Reaction Space in porous Cermet Anodes based on Ion-conducting Electrolytes: Patterns of Degradation. *Solid State Ionics*, 124(3-4):221 – 237, 1999.

[55] T. Matsui, R. Kishida, J.-Y. Kim, H. Muroyama, and K. Eguchi. Performance Deterioration of Ni-YSZ Anode Induced by Electrochemically Generated Steam in Solid Oxide Fuel Cells. *Journal of The Electrochemical Society*, 157(5):B776–B781, 2010.

[56] S. Koch, P. V. Hendriksen, M. Mogensen, Y.-L. Liu, N. Dekker, B. Rietveld, B. de Haart, and F. Tietz. Solid Oxide Fuel Cell Performance under Severe Operating Conditions. *Fuel Cells*, 6(2):130–136, 2006.

[57] R. Vaßen, D. Simwonis, and D. Stöver. Modelling of the Agglomeration of Ni-Particles in Anodes of Solid Oxide Fuel Cells. *Journal of Materials Science*, 36(1):147–151, 2001.

[58] D. Simwonis, F. Tietz, and D. Stöver. Nickel Coarsening in Annealed Ni/8YSZ Anode Substrates for Solid Oxide Fuel Cells. *Solid State Ionics*, 132(3-4):241 – 251, 2000.

[59] E. Ivers-Tiffée, H. Timmermann, A. Leonide, N. H. Menzler, and J. Malzbender. Methane Reforming Kinetics, Carbon Deposition, and Redox Durability of Ni/8 Yttria-stabilized Zirconia (YSZ) Anodes. In *Handbook of Fuel Cells*, chapter 64. John Wiley & Sons, Ltd, 2010.

[60] Q. Fu, F. Tietz, D. Sebold, Sh. Tao, and J. T. S. Irvine. An Efficient Ceramic-based Anode for Solid Oxide Fuel Cells. *Journal of Power Sources*, 171(2):663 – 669, 2007.

[61] M. D. Gross, J. M. Vohs, and R. J. Gorte. A Strategy for Achieving High Performance with SOFC Ceramic Anodes. *Electrochemical and Solid-State Letters*, 10(4):B65–B69, 2007.

[62] B. Butz. *Yttria-Doped Zirconia as Solid Electrolyte for Fuel-Cell Applications*. PhD thesis, Karlsruher Institut für Technologie (KIT), 2009.

[63] C. Endler-Schuck. *Alterungsverhalten mischleitender LSCF Kathoden für Hochtemperatur-Festoxid-Brennstoffzellen (SOFCs)*. PhD thesis, Karlsruher Institut für Technologie (KIT), 2012. ; Pb.: EUR 36,00.

[64] A. Mai, V. A. C. Haanappel, F. Tietz, and D. Stöver. Ferrite-based Perovskites as Cathode Materials for Anode-supported Solid Oxide Fuel Cells: Part II. Influence of the CGO Interlayer. *Solid State Ionics*, 177(19-25):2103 – 2107, 2006.

[65] S. Uhlenbruck, N. Jordan, D. Sebold, H. P. Buchkremer, V. A. C. Haanappel, and D. Stöver. Thin film coating technologies of (ce,gd)o2-d interlayers for application in ceramic high-temperature fuel cells. *Thin Solid Films*, 515(7-8):4053 – 4060, 2007.

[66] C. Endler-Schuck, A. Leonide, A. Weber, S. Uhlenbruck, F. Tietz, and E. Ivers-Tiffée. Performance Analysis of Mixed Ionic-Electronic Conducting Cathodes in Anode Supported Cells. *Journal of Power Sources*, 196(17):7257 – 7262, 2011.

[67] M. Kornely. *Elektrische Charakterisierung und Modellierung von metallischen Interkonnektoren (MIC) des SOFC-Stacks*. PhD thesis, Karlsruher Institut für Technologie (KIT), 2012.

[68] A. Mai, V. A. C. Haanappel, S. Uhlenbruck, F. Tietz, and D. Stöver. Ferrite-based Perovskites as Cathode Materials for Anode-supported Solid Oxide Fuel Cells: Part I. Variation of Composition. *Solid State Ionics*, 176(15-16):1341 – 1350, 2005.

[69] L. Blum, H.-P. Buchkremer, S. M. Gross, B. De Haart, J. W. Quadakkers, U. Reisgen, R. Steinberger-Wilckens, R. W. Steinbrech, and F. Tietz. Overview of the Development of Solid Oxide Fuel Cells at Forschungszentrum Jülich. *Proceedings - Electrochemical Society*, PV 2005-07:39–47, 2005.

[70] L. Blum, S. M. Groß, J. Malzbender, U. Pabst, M. Peksen, R. Peters, and I. C. Vinke. Investigation of Solid Oxide Fuel Cell Sealing Behavior under Stack relevant Conditions at Forschungszentrum Jülich. *Journal of Power Sources*, 196(17):7175 – 7181, 2011.

[71] M. Winter and K.-C. Möller. Primäre und wiederaufladbare Lithium-Batterien. Technical report, Institut für Chemische Technologie Anorganischer Stoffe, TU Graz, 2005. Script zum Praktikum Anorganisch-Chemische Technologie.

[72] A. Jossen and W. Weydanz. *Moderne Akkumulatoren richtig einsetzen*. Ubooks, Neusäß, 1. edition, 2006.

[73] B. Frenzel, P. Kurzweil, and H. Rönnebeck. Electromobility Concept for Racing Cars based on Lithium-Ion Batteries and Supercapacitors. *Journal of Power Sources*, 196(12):5364 – 5376, 2011.

[74] M. Ender, J. Joos, T. Carraro, and E. Ivers-Tiffée. Three-Dimensional Reconstruction of a Composite Cathode for Lithium-Ion Cells. *Electrochemistry Communications*, 13(2):166 – 168, 2011.

[75] A. Utz. *The Electrochemical Oxidation of H_2 and CO at Patterned Ni Anodes of SOFCs*. PhD thesis, Karlsruher Institut für Technologie (KIT), Karlsruhe, 2011.

[76] J. Van herle, N. Autissier, D. Favrat, D. Larrain, Z. Wuillemin, and M. Molinelli. Modeling and Experimental Validation of Solid Oxide Fuel Cell Materials and Stacks. *Journal of the European Ceramic Society*, 25(12 spec. iss.):2627–2632, 2005.

[77] A. C. Müller. *Mehrschicht-Anode für die Hochtemperatur-Brennstoffzelle (SOFC)*. PhD thesis, Universität Karlsruhe, 2005.

[78] M. Pihlatie, A. Kaiser, and M. Mogensen. Redox Stability of SOFC: Thermal Analysis of Ni-YSZ Composites. *Solid State Ionics*, 180(17-19):1100 – 1112, 2009.

[79] A. Weber, B. Rüger, D. Klotz, and M. Kornely. Herausforderung Brennstoffzelle: 'Betrieb von Hochtemperatur-Brennstoffzellen (SOFC) unter transienten Lastbedingungen'. Technical report, Institut für Werkstoffe der Elektrotechnik, 2009.

[80] W. Weppner. Voltage Relaxation Measurements of the Electron and Hole Mobilities in Yttria-doped Zirconia. *Electrochimica Acta*, 22(7):721 – 727, 1977.

[81] D. Klotz, A. Leonide, and E. Ivers-Tiffée. Recovery of Anode Performance by Reverse Current Treatment. *ECS Transactions*, 25(2):2049–2056, 2009.

[82] D. Klotz, B. Butz, A. Leonide, D. Gerthsen, and E. Ivers-Tiffée. Increase of Anode Performance of SOFC by Reverse Current Treatment. *ECS Transactions*, 28(11):141–150, 2010.

[83] D. Klotz, B. Butz, A. Leonide, J. Hayd, D. Gerthsen, and E. Ivers-Tiffée. Performance Enhancement of SOFC Anode through Electrochemically Induced Ni/YSZ Nanostructures. *Journal of The Electrochemical Society*, 158(6):B587–B595, 2011.

[84] D. Klotz, J. Szasz, A. Weber, and E. Ivers-Tiffée. Nano-Structuring of SOFC Anodes by Reverse Current Treatment. *ECS Transactions*, 45(1):241–249, 2012.

[85] J. Janek and C. Korte. Electrochemical Blackening of Yttria-stabilized Zirconia - Morphological Instability of the Moving Reaction Front. *Solid State Ionics*, 116(3-4):181 – 195, 1999.

[86] J. Schefold, A. Brisse, and M. Zahid. Electronic Conduction of Yttria-stabilized Zirconia Electrolyte in Solid Oxide Cells Operated in High Temperature Water Electrolysis. *Journal of The Electrochemical Society*, 156(8):B897–B904, 2009.

[87] W. Weppner. Electronic Transport Properties and Electrically Induced p-n Junction in ZrO_2 + 10 mol% Y_2O_3. *Journal of Solid State Chemistry*, 20(3):305 – 314, 1977.

[88] S. S. Zhang, K. Xu, and T. R. Jow. Study of the Charging Process of a LiCoO$_2$-based Li-Ion Battery. *Journal of Power Sources*, 160(2):1349 – 1354, 2006.

[89] K. Eberman, P. Gomadam, G. Jain, and E. Scott. Material and Design Options for Avoiding Lithium Plating during Charging. *ECS Transactions*, 25(35):47–58, 2010.

[90] T. Ohzuku, S. Takeda, and M. Iwanaga. Solid-State Redox Potentials for Li[Me$_{1/2}$Mn$_{3/2}$]O$_4$ (Me: 3d-Transition Metal) having Spinel-Framework Structures: a Series of 5 Volt Materials for Advanced Lithium-Ion Batteries. *Journal of Power Sources*, 81-82(0):90 – 94, 1999.

[91] C. Endler, A. Leonide, A. Weber, F. Tietz, and E. Ivers-Tiffée. Time-Dependent Electrode Performance Changes in Intermediate Temperature Solid Oxide Fuel Cells. *Journal of The Electrochemical Society*, 157(2):B292–B298, 2010.

[92] A. Leonide, S. Ngo Dinh, A. Weber, and E. Ivers-Tiffée. Performance Limiting Factors in Anode Supported SOFC. *Proceedings of the 8th European Solid Oxide Fuel Cell Forum*, page A0501, 2008.

[93] L. Ljung. Perspectives on System Identification. *Annual Reviews in Control*, 34(1):1 – 12, 2010.

[94] R. Bove, editor. *Modeling Solid Oxide Fuel Cells : Methods, Procedures and Techniques*. Fuel cells and hydrogen energy ; 1. Springer, Dordrecht, 2008.

[95] W. G. Bessler, S. Gewies, and M. Vogler. A New Framework for Physically based Modeling of Solid Oxide Fuel Cells. *Electrochimica Acta*, 53(4):1782 – 1800, 2007.

[96] V. Yurkiv, A. Utz, A. Weber, E. Ivers-Tiffée, H.-R. Volpp, and W. G. Bessler. Elementary Kinetic Modeling and Experimental Validation of Electrochemical CO Oxidation on Ni/YSZ Pattern Anodes. *Electrochimica Acta*, 59(0):573 – 580, 2012.

[97] B. Rüger, A. Weber, D. Fouquet, and E. Ivers-Tiffée. SOFC Performance Evaluated by FEM-Modelling of Electrode Microstructures with a Randomly Generated Geometry. *Proceedings of the 7th European Solid Oxide Fuel Cell Forum*, page B075, 2006.

[98] J. Joos, M. Ender, T. Carraro, A. Weber, and E. Ivers-Tiffée. Representative Volume Element Size for Accurate Solid Oxide Fuel Cell Cathode Reconstructions from Focused Ion Beam Tomography Data. *Electrochimica Acta*, (0):–, 2012. accepted manuscript.

[99] J. Joos, T. Carraro, M. Ender, B. Rüger, A. Weber, and E. Ivers-Tiffée. Detailed Microstructure Analysis and 3D Simulations of Porous Electrodes. *ECS Transactions*, 35(1):2357–2368, 2011.

[100] A. Häffelin, J. Joos, M. Ender, A. Weber, and E. Ivers-Tiffée. Transient 3D FEM Impedance-Model for Mixed Conducting Cathodes. *ECS Transactions*, 45(1):313–325, 2012.

[101] L. Andreassi, S. Ubertini, R. Bove, and N. M. Sammes. Cfd-based results for planar and micro-tubular single cell designs. In Roberto Bove and Stefano Ubertini, editors, *Modeling Solid Oxide Fuel Cells*, Fuel Cells and Hydrogen Energy, pages 97–122. Springer Netherlands, 2008.

[102] N. Autissier, D. Larrain, J. Van herle, and D. Favrat. CFD Simulation Tool for Solid Oxide Fuel Cells. *Journal of Power Sources*, 131(1-2):313 – 319, 2004.

[103] M. Doyle, T. F. Fuller, and J. Newman. Modeling of Galvanostatic Charge and Discharge of the Lithium/Polymer/Insertion Cell. *Journal of The Electrochemical Society*, 140(6):1526–1533, 1993.

[104] A. Leonide, Y. Apel, and E. Ivers-Tiffee. SOFC Modeling and Parameter Identification by Means of Impedance Spectroscopy. *ECS Transactions*, 19(20):81–109, 2009.

[105] M. P. Eschenbach, R. Coulon, A. A. Franco, J. Kallo, and W. G. Bessler. Multi-Scale Simulation of Fuel Cells: From the Cell to the System. *Solid State Ionics*, 192(1):615 – 618, 2011.

[106] J. Lunze. *Künstliche Intelligenz für Ingenieure*, volume 1. Oldenbourg, München, 1994.

[107] J. Lunze. *Künstliche Intelligenz für Ingenieure*, volume 2. Oldenbourg, München, 1995.

[108] M. Buchholz. *Subspace-Identification zur Modellierung von PEM-Brennstoffzellen-Stacks*. PhD thesis, Karlsruher Institut für Technologie (KIT), 2010.

[109] P. Berg. *Modellierung einer Lithium-Ionen-Batterie mittels eines linearen, zeitvarianten Systemansatzes*. Bachelor thesis, Karlsruher Institut für Technologie (KIT), 2011.

[110] D. Klotz, M. Schönleber, J. P. Schmidt, and E. Ivers-Tiffée. New Aapproach for the Calculation of Impedance Spectra out of Time Domain Data. *Electrochimica Acta*, 56(24):8763 – 8769, 2011.

[111] H. Alagi. *Bestimmung der Reaktionsentropie von Li-Ionen Zellen*. Bachelor thesis, Karlsruher Institut für Technologie (KIT), 2011.

[112] A. Weber. *Entwicklung von Kathodenstrukturen für die Hochtemperatur-Brennstoffzelle SOFC*. PhD thesis, Universität Karlsruhe, 2003.

[113] M. Becker, A. Mai, E. Ivers-Tiffée, and F. Tietz. Long-Term Measurements of Anode-supported Solid Oxide Fuel Cells with LSCF Cathode under various Operating Conditions. *Proceedings of the Electrochemical Society*, 7:514–523, 2005.

[114] M. Ender, J. Joos, T. Carraro, and E. Ivers-Tiffée. Three-Dimensional Reconstruction of a Composite Cathode for Lithium-Ion Cells. *Electrochemistry Communications*, 13(2):166 – 168, 2011.

[115] M. S. Naidu and V. Kamaraju. *High Voltage Engineering*. A McGraw-Hill special reprint edition. McGraw-Hill, New York [u.a.], 2. edition, 1996.

[116] M. Dubarry, V. Svoboda, R. Hwu, and B. Y. Liaw. Incremental Capacity Analysis and Close-to-Equilibrium OCV Measurements to Quantify Capacity Fade in Commercial Rechargeable Lithium Batteries. *Electrochemical and Solid-State Letters*, 9(10):A454–A457, 2006.

[117] A. Tranchant, J. M. Blengino, J. Farcy, and R. Messina. Study of the Lithium Intercalation/Deintercalation Processes into V_2O_5 by Linear Voltammetry at Slow Sweep Rates. *Journal of The Electrochemical Society*, 139(5):1243–1248, 1992.

[118] J. Richter. *Messung von quasistationären Kennlinien von Batterien und die Simulation der Leerlaufspannung*. Diploma thesis, Karlsruher Institut für Technologie (KIT), 2011.

[119] I. Bloom, A. N. Jansen, D. P. Abraham, J. Knuth, S. A. Jones, V. S. Battaglia, and G. L. Henriksen. Differential Voltage Analyses of High-Power, Lithium-Ion Cells: 1. Technique and Application. *Journal of Power Sources*, 139(1-2):295 – 303, 2005.

[120] K. West, T. Jacobsen, B. Zachau-Christiansen, and S. Atlung. Determination of the Differential Capacity of Intercalation Electrode Materials by Slow Potential Scans. *Electrochimica Acta*, 28(1):97 – 107, 1983.

[121] D. D. Macdonald. Reflections on the Hhistory of Electrochemical Impedance Spectroscopy. *Electrochimica Acta*, 51(8-9):1376 – 1388, 2006.

[122] B. A. Boukamp. Practical Application of the Kramers-Kronig Transformation on Impedance Measurements in Solid State Electrochemistry. *Solid State Ionics*, 62(1-2):131 – 141, 1993.

[123] Servowatt GmbH. personal communication. Servowatt is a company offering a wide range of high frequency potentiostats, 2008.

[124] V. Sonn. *Fehlerquellen der Charakterisierung niederimpedanter Festelektrolyt Brennstoffzellen (SOFC)*. Study project, Universität Karlsruhe, 2011.

[125] E. Van Gheem, J. Vereecken, J. Schoukens, R. Pintelon, P. Guillaume, P. Verboven, and L. Pauwels. Instantaneous Impedance Measurements on Aluminium using a Schroeder Multisine Excitation Signal. *Electrochimica Acta*, 49(17-18):2919 – 2925, 2004.

[126] Y. Van Ingelgem, E. Tourwé, O. Blajiev, R. Pintelon, and A. Hubin. Advantages of Odd Random Phase Multisine Electrochemical Impedance Measurements. *Electroanalysis*, 21(6):730–739, 2009.

[127] K. Darowicki, J. Orlikowski, and G. Lentka. Instantaneous Impedance Spectra of a Non-Stationary Model Electrical System. *Journal of Electroanalytical Chemistry*, 486(2):106 – 110, 2000.

[128] E. Barsoukov, S. H. R., and H. Lee. A Novel Impedance Spectrometer based on Carrier Function Laplace-Transform of the Response to Arbitrary Excitation. *Journal of Electroanalytical Chemistry*, 536(1-2):109 – 122, 2002.

[129] S. Hansmann. *Modellierung und Simulation des Spannungsverhaltens einer Lithium-Ionen-Zelle.* Diploma thesis, Karlsruher Institut für Technologie (KIT), 2011.

[130] J. T. Zhang, J. M. Hu, J. Q. Zhang, and C. N. Cao. Studies of Impedance Models and Water Transport Behaviors of Polypropylene Coated Metals in NaCl Solution. *Progress in Organic Coatings*, 49(4):293 – 301, 2004.

[131] B. R. Hinderliter, K. N. Allahar, G. P. Bierwagen, D. E. Tallman, and S. G. Croll. Thermal Cycling of Epoxy Coatings Using Room Temperature Ionic Liquids. *Journal of The Electrochemical Society*, 155(3):C93–C100, 2008.

[132] *EIS300 Electrochemical Impedance Spectroscopy Software.* www.gamry.com. EIS300-Product-Brochure.

[133] D. Klotz, A. Weber, and E. Ivers-Tiffée. Thermal Behaviour of SOFC Single Cells in Operation with Dynamic Loads. *Proceedings of the 8th European SOFC Forum*, 1:B0909, 2008.

[134] H. Köhler. *Analyse des dynamischen Temperaturverhaltens der SOFC unter elektrischer Last.* Study project, Karlsruher Institut für Technologie (KIT), 2009.

[135] J. R. Wilson, D. T. Schwartz, and S. B. Adler. Nonlinear Electrochemical Impedance Spectroscopy for Solid Oxide Fuel Cell Cathode Materials. *Electrochimica Acta*, 51(8-9):1389 – 1402, 2006.

[136] J. R. Wilson, M. Sase, T. Kawada, and S. B. Adler. Measurement of Oxygen Exchange Kinetics on Thin-Film $La_{0.6}Sr_{0.4}CoO_{3-\delta}$ Using Nonlinear Electrochemical Impedance Spectroscopy. *Electrochemical and Solid-State Letters*, 10(5):B81–B86, 2007.

[137] L. Ljung. *System identification : theory for the user.* Prentice-Hall information and system sciences series. Prentice Hall, Upper Saddle River, NJ [u.a.], 2. edition, 1999.

[138] R. Land, P. Annus, and M. Min. Time-Frequency Impedance Spectroscopy: Excitation Considerations. In *Proc. of the 15th IMEKO TC4 International Symposium on Novelties in Electrical Measurements and Instrumentations*, Iasi, Romania, 2007.

[139] O. Föllinger. *Laplace-, Fourier- und z-Transformation.* Hüthig, Heidelberg, 7. edition, 2000.

[140] G. S. Popkirov and R. N. Schindler. Validation of Experimental Data in Electrochemical Impedance Spectroscopy. *Electrochimica Acta*, 38(7):861 – 867, 1993.

[141] J. E. Garland, C. M. Pettit, and D. Roy. Analysis of Experimental Constraints and Variables for Time Resolved Detection of Fourier Transform Electrochemical Impedance Spectra. *Electrochimica Acta*, 49(16):2623 – 2635, 2004.

[142] A. A. Pilla. A Transient Impedance Technique for the Study of Electrode Kinetics. *Journal of The Electrochemical Society*, 117(4):467–477, 1970.

[143] K. Takano, K. Nozaki, Y. Saito, K. Kato, and A. Negishi. Impedance Spectroscopy by Voltage-Step Chronoamperometry Using the Laplace Transform Method in a

Lithium-Ion Battery. *Journal of The Electrochemical Society*, 147(3):922–929, 2000.

[144] U. Kiencke and H. Jäkel. *Signale und Systeme*. Oldenbourg, München, 3. edition, 2005.

[145] M. Schönleber. *Entwicklung eines kombinierten Messverfahrens im Zeit- und Frequenzbereich für Lithium-Ionen-Batterien*. Study project, Karlsruher Institut für Technologie (KIT), 2010.

[146] W. J. Thompson and J. R. Macdonald. Discrete and Integral Fourier Transforms: Analytical Examples. *Proceedings of the National Academy of Sciences of the United States of America*, 90(15):pp. 6904–6908, 1993.

[147] H. Unbehauen. *Regelungstechnik*, volume 1: Klassische Verfahren zur Analyse und Synthese linearer kontinuierlicher Regelsysteme, Fuzzy-Regelsysteme. Vieweg, Braunschweig, 13. edition, 2005.

[148] B. A. Boukamp. Electrochemical Impedance Spectroscopy in Solid State Ionics: Recent Advances. *Solid State Ionics*, 169(1-4):65 – 73, 2004.

[149] G. S. Popkirov and R. N. Schindler. A New Impedance Spectrometer for the Investigation of Electrochemical Systems. *Review of Scientific Instruments*, 63(11):5366–5372, 1992.

[150] J. R. J. Larminie. Current Interrupt Techniques for Circuit Modelling. In *IEE Colloquium on Electrochemical Measurement*, pages 12/1 –12/6, 1994.

[151] F. Richter, C.-A. Schiller, and N. Wagner. *Current Interrupt Technique - Measuring Low Impedances at High Frequencies*. Zahner-elektrik GmbH & Co. KG, electrochemical applications edition, 2002. www.zahner.de.

[152] L. Schindele. *Einsatz eines leistungselektronischen Stellglieds zur Parameteridentifikation und optimalen Betriebsführung von PEM-Brennstoffzellensystemen*. PhD thesis, Universität Karlsruhe, 2006.

[153] W. Weppner and R. A. Huggins. Determination of the Kinetic Parameters of Mixed-Conducting Electrodes and Application to the System Li_3Sb. *Journal of The Electrochemical Society*, 124(10):1569–1578, 1977.

[154] C. J. Wen, B. A. Boukamp, R. A. Huggins, and W. Weppner. Thermodynamic and Mass Transport Properties of "LiAl". *Journal of The Electrochemical Society*, 126(12):2258–2266, 1979.

[155] H.-R. Tränkler, editor. *Sensortechnik : Handbuch für Praxis und Wissenschaft*. Springer, Berlin, 1998.

[156] D. Klotz, J. P. Schmidt, A. Kromp, A. Weber, and E. Ivers-Tiffée. The Distribution of Relaxation Times as Beneficial Tool for Equivalent Circuit Modeling of Fuel Cells and Batteries. *ECS Transactions*, 41(28):25–33, 2012.

[157] B. A. Boukamp. A Linear Kronig-Kramers Transform Test for Immittance Data Validation. *Journal of The Electrochemical Society*, 142(6):1885–1894, 1995.

[158] B. A. Boukamp and J. R. Macdonald. Alternatives to Kronig-Kramers Transformation and Testing, and Estimation of Distributions. *Solid State Ionics*, 74(1-2):85 – 101, 1994.

[159] H. Schichlein, A. C. Müller, M. Voigts, A. Krügel, and E. Ivers-Tiffée. Deconvolution of Electrochemical Iimpedance Spectra for the Identification of Electrode Reaction Mechanisms in Solid Oxide Fuel Cells. *Journal of Applied Electrochemistry*, 32:875–882, 2002.

[160] J. Weese. A Reliable and Fast Method for the Solution of Fredhol Integral Equations of the First Kind based on Tikhonov Regularization. *Computer Physics Communications*, 69(1):99 – 111, 1992.

[161] A. N. Tichonov, editor. *Numerical Methods for the Solution of Ill-posed Problems.* Mathematics and its Applications. Kluwer Academic, Dordrecht [u.a.], 1995.

[162] V. Sonn, A. Leonide, and E. Ivers-Tiffée. Combined Deconvolution and CNLS Fitting Approach Applied on the Impedance Response of Technical Ni/8YSZ Cermet Electrodes. *Journal of The Electrochemical Society*, 155(7):B675–B679, 2008.

[163] D. Johnson. *Zplot, ZView Electrochemical Impedence Software, version 2.3b.* Scriber Associates Inc., Southern Pines, NC, USA, 2000.

[164] J. J. Moré. The Levenberg-Marquardt Algorithm: Implementation and Theory. In *Numerical analysis (Proc. 7th Biennial Conf., Univ. Dundee, Dundee, 1977)*, pages 105–116. Lecture Notes in Math., Vol. 630. Springer, Berlin, 1978.

[165] K. Levenberg. A Method for the Solution of Certain Problems in Least Squares. *Quart. Applied Math.*, 2:164–168, 1944.

[166] D. W. Marquardt. An Algorithm for Least-Squares Estimation of Nonlinear Parameters. *SIAM Journal on Applied Mathematics*, 11(2):431–441, 1963.

[167] J. C. Lagarias, J. A. Reeds, M. H. Wright, and P. E. Wright. Convergence Properties of the Nelder-Mead Simplex Method in Low Dimensions. *SIAM JOURNAL ON OPTIMIZATION*, 9(1):112–147, DEC 21 1998.

[168] J. A. Nelder and R. Mead. A Simplex Method for Function Minimization. *The Computer Journal*, 7(4):308–313, 1965.

[169] Q.-A. Huang, R. Hui, B. Wang, and J. Zhang. A Review of AC Impedance Modeling and Validation in SOFC Diagnosis. *Electrochimica Acta*, 52(28):8144 – 8164, 2007.

[170] S. H. Jensen, A. Hauch, P. V. Hendriksen, M. Mogensen, N. Bonanos, and T. Jacobsen. A Method to Separate Process Contributions in Impedance Spectra by Variation of Test Conditions. *Journal of The Electrochemical Society*, 154(12):B1325–B1330, 2007.

[171] B. Liu, T. Matsui, H. Muroyama, K. Tomida, T. Kabata, and K. Eguchi. Impedance Analysis of Practical Segmented-in-Series Tubular Solid Oxide Fuel Cells. *ECS Transactions*, 35(1):637–646, 2011.

[172] W. G. Bessler. Gas Concentration Impedance of Solid Oxide Fuel Cell Anodes. *Journal of The Electrochemical Society*, 153(8):A1492–A1504, 2006.

[173] D. Klotz, A. Weber, and E. Ivers-Tiffée. Dynamic Electrochemical Model for SOFC Stacks. *ECS Transactions*, 25(2):1234–5678, 2009.

[174] S. Singhal. Advances in Solid Oxide Fuel Cell Technology. *Solid State Ionics*, 135(1-4):305 – 313, 2000.

[175] W. Winkler and P. Nehter. Thermodynamics of Fuel Cells. In Roberto Bove and Stefano Ubertini, editors, *Modeling Solid Oxide Fuel Cells*, Fuel Cells and Hydrogen Energy, pages 13–50. Springer Netherlands, 2008.

[176] J. Stahl. *Modellierung des stationären Verhaltens einer Feststoff-Elektrolysezelle (SOEC)*. Study project, Karlsruher Institut für Technologie (KIT), 2012.

[177] A. Leonide, B. Rüger, A. Weber, W. A. Meulenberg, and E. Ivers-Tiffée. Impedance Study of Alternative $(La,Sr)FeO_{3-\delta}$ and $(La,Sr)(Co,Fe)O_{3-\delta}$ MIEC Cathode Compositions. *Journal of The Electrochemical Society*, 157(2):B234–B239, 2010.

[178] J. P. Schmidt. *Entwicklung eines Diagnose Tools für eine SOFC APU*. Diploma thesis, Karlsruher Institut für Technologie (KIT), 2009.

[179] A. Leonide, S. Hansmann, and E. Ivers-Tiffée. A 0-Dimensional Stationary Model for Anode-Supported Solid Oxide Fuel Cells. *ECS Transactions*, 28(11):341–346, 2010.

[180] K. Schmid. *Qualitative dynamische Modelle für das Verhalten von Hochtemperatur-Brennstoffzellen*. PhD thesis, Universität Karlsruhe, Düsseldorf, 2003.

[181] A. Bieberle and L. J. Gauckler. State-Space Modeling of the Anodic SOFC System. *Solid State Ionics*, 146(1-2):23 – 41, 2002.

[182] H. Zhu and R. J. Kee. Modeling Electrochemical Impedance Spectra in SOFC Button Cells with Internal Methane Reforming. *Journal of The Electrochemical Society*, 153(9):A1765–A1772, 2006.

[183] H. Geisler. *Modellierung und transiente Simulation der Stofftransportvorgänge im Anodensubstrat der Hochtemperatur-Brennstoffzelle*. Diploma thesis, Karlsruher Institut für Technologie (KIT), 2011.

[184] M. Pfafferodt, P. Heidebrecht, M. Stelter, and K. Sundmacher. Model-based Prediction of Suitable Operating Range of a SOFC for an Auxiliary Power Unit. *Journal of Power Sources*, 149(0):53 – 62, 2005.

[185] C.-K. Lin, T.-T. Chen, Y.-P. Chyou, and L.-K. Chiang. Thermal Stress Analysis of a Planar SOFC Stack. *Journal of Power Sources*, 164(1):238 – 251, 2007.

[186] C.-K. Lin, L.-H. Huang, L.-K. Chiang, and Y.-P. Chyou. Thermal Stress Analysis of Planar Solid Oxide Fuel Cell Stacks: Effects of Sealing Design. *Journal of Power Sources*, 192(2):515 – 524, 2009.

[187] B. Liu, H. Muroyama, T. Matsui, K. Tomida, T. Kabata, and K. Eguchi. Gas Transport Impedance in Segmented-in-Series Tubular Solid Oxide Fuel Cell. *Journal of The Electrochemical Society*, 158(2):B215–B224, 2011.

[188] A. Momma, Y. Kaga, K. Takano, K. Nozaki, A. Negishi, K. Kato, T. Kato, T. Inagaki, H. Yoshida, K. Hosoi, K. Hoshino, T. Akbay, J. Akikusa, M. Yamada, and N. Chitose. AC Impedance Behavior of a Practical-Size Single-Cell SOFC under DC Current. *Solid State Ionics*, 174(1-4):87 – 95, 2004.

[189] S. Primdahl and M. Mogensen. Gas Conversion Impedance: A Test Geometry Effect in Characterization of Solid Oxide Fuel Cell Anodes. *Journal of The Electrochemical Society*, 145(7):2431–2438, 1998.

[190] P. V. Hendriksen, S. Koch, M. Mogensen, Y. L. Liu, and P. H. Larsen. Breakdown of Losses in Thin Electrolyte SOFCs. *Proceedings - Electrochemical Society*, PV 2003-07:1147–1157, 2003.

Werkstoffwissenschaft @ Elektrotechnik /

Universität Karlsruhe, Institut für Werkstoffe der Elektrotechnik

Die Bände sind im Verlagshaus Mainz (Aachen) erschienen.

Band 1   Helge Schichlein
          Experimentelle Modellbildung für die Hochtemperatur-Brennstoff-
          zelle SOFC. 2003
          ISBN 3-86130-229-2

Band 2   Dirk Herbstritt
          Entwicklung und Optimierung einer leistungsfähigen Kathoden-
          struktur für die Hochtemperatur-Brennstoffzelle SOFC. 2003
          ISBN 3-86130-230-6

Band 3   Frédéric Zimmermann
          Steuerbare Mikrowellendielektrika aus ferroelektrischen
          Dickschichten. 2003
          ISBN 3-86130-231-4

Band 4   Barbara Hippauf
          Kinetik von selbsttragenden, offenporösen Sauerstoffsensoren
          auf der Basis von $Sr(Ti,Fe)O_3$. 2005
          ISBN 3-86130-232-2

Band 5   Daniel Fouquet
          Einsatz von Kohlenwasserstoffen in der Hochtemperatur-Brennstoff-
          zelle SOFC. 2005
          ISBN 3-86130-233-0

Band 6   Volker Fischer
          Nanoskalige Nioboxidschichten für den Einsatz in hochkapazitiven
          Niob-Elektrolytkondensatoren. 2005
          ISBN 3-86130-234-9

Band 7   Thomas Schneider
          Strontiumtitanferrit-Abgassensoren.
          Stabilitätsgrenzen / Betriebsfelder. 2005
          ISBN 3-86130-235-7

Band 8   Markus J. Heneka
          Alterung der Festelektrolyt-Brennstoffzelle unter thermischen
          und elektrischen Lastwechseln. 2006
          ISBN 3-86130-236-5

Band 9   Thilo Hilpert
          Elektrische Charakterisierung von Wärmedämmschichten mittels
          Impedanzspektroskopie. 2007
          ISBN 3-86130-237-3

Band 10   Michael Becker
           Parameterstudie zur Langzeitbeständigkeit von Hochtemperatur-
           brennstoffzellen (SOFC). 2007
           ISBN 3-86130-239-X

Band 11   Jin Xu
           Nonlinear Dielectric Thin Films for Tunable Microwave Applications.
           2007
           ISBN 3-86130-238-1

Band 12   Patrick König
           Modellgestützte Analyse und Simulation von stationären Brennstoff-
           zellensystemen. 2007
           ISBN 3-86130-241-1

Band 13   Steffen Eccarius
           Approaches to Passive Operation of a Direct Methanol Fuel Cell. 2007
           ISBN 3-86130-242-X

Fortführung als

Schriften des Instituts für Werkstoffe der Elektrotechnik, Karlsruher Institut für Technologie (1868-1603)

bei KIT Scientific Publishing

Die Bände sind unter www.ksp.kit.edu als PDF frei verfügbar oder
als Druckausgabe bestellbar.

Band 14   Stefan F. Wagner
           Untersuchungen zur Kinetik des Sauerstoffaustauschs an
           modifizierten Perowskitgrenzflächen. 2009
           ISBN 978-3-86644-362-4

Band 15   Christoph Peters
           Grain-Size Effects in Nanoscaled Electrolyte and Cathode Thin Films
           for Solid Oxide Fuel Cells (SOFC). 2009
           ISBN 978-3-86644-336-5

Band 16   Bernd Rüger
           Mikrostrukturmodellierung von Elektroden für die Festelektro-
           lytbrennstoffzelle. 2009
           ISBN 978-3-86644-409-6

Band 17   Henrik Timmermann
           Untersuchungen zum Einsatz von Reformat aus flüssigen Kohlen-
           wasserstoffen in der Hochtemperaturbrennstoffzelle SOFC. 2010
           ISBN 978-3-86644-478-2